AF769043

# Recent Advances in Phytochemical Research

*Edited by Muhammad Kamran Khan*

Published in London, United Kingdom

Recent Advances in Phytochemical Research
http://dx.doi.org/10.5772/10.5772/intechopen.1003331
Edited by Muhammad Kamran Khan

Contributors
Abdullah Alsrhani, Aisha Farhana, Ana Montserrat Corona-España, Aparna A. Saraf, Ayman A.M. Alameen, Dexter Achu Mosoh, Dinh Thanh Giang, Divya Lobo Padinjarekutt, Dwijalee Deo, Elijah Kipkorir Lelmen, Emad Manni, Ingrid Mayanin Rodriguez-Buenfil, Jacqueline K. Makatiani, Lienda Bashier Eltayeb, Liviya Gaikwad, Mario Alberto Garcia-Ramirez, Mridula Saikia Barooah, Muhammad Atif, Nada Alhathlaul, Neha Mourya, Nergiz Gürbüz Çolak, Nguyen Minh Duc, Nguyen Van Khiem, Nikki Huria, Okram Abemsana Devi, Orfil González Reynoso, Regina Romo-Gonzalez, Shubham Burande, Shubham V.U. Tanpathak, Wassila Derafa, Yusuf Saleem Khan

© The Editor(s) and the Author(s) 2025
The rights of the editor(s) and the author(s) have been asserted in accordance with the Copyright, Designs and Patents Act 1988. All rights to the book as a whole are reserved by INTECHOPEN LIMITED. The book as a whole (compilation) cannot be reproduced, distributed or used for commercial or non-commercial purposes without INTECHOPEN LIMITED's written permission. Enquiries concerning the use of the book should be directed to INTECHOPEN LIMITED rights and permissions department (permissions@intechopen.com).
Violations are liable to prosecution under the governing Copyright Law.

Individual chapters of this publication are distributed under the terms of the Creative Commons Attribution 4.0 License which permits commercial use, distribution and reproduction of the individual chapters, provided the original author(s) and source publication are appropriately acknowledged. If so indicated, certain images may not be included under the Creative Commons license. In such cases users will need to obtain permission from the license holder to reproduce the material. More details and guidelines concerning content reuse and adaptation can be found at http://www.intechopen.com/copyright-policy.html.

Notice
Statements and opinions expressed in the chapters are these of the individual contributors and not necessarily those of the editors or publisher. No responsibility is accepted for the accuracy of information contained in the published chapters. The publisher assumes no responsibility for any damage or injury to persons or property arising out of the use of any materials, instructions, methods or ideas contained in the book.

First published in London, United Kingdom, 2025 by IntechOpen
IntechOpen is the global imprint of INTECHOPEN LIMITED, registered in England and Wales, registration number: 11086078, 167-169 Great Portland Street, London, W1W 5PF, United Kingdom

For EU product safety concerns: IN TECH d.o.o., Prolaz Marije Krucifikse Kozulić 3, 51000 Rijeka, Croatia, info@intechopen.com or visit our website at intechopen.com.

British Library Cataloguing-in-Publication Data
A catalogue record for this book is available from the British Library

Recent Advances in Phytochemical Research
Edited by Muhammad Kamran Khan
p. cm.

This title is part of the Biochemistry Book Series, Volume 64
Topic: Chemical Biology
Series Editor: Andrei Surguchov
Topic Editors: Murat Şentürk and Deniz Ekinci

Print ISBN 978-0-85014-838-1
Online ISBN 978-0-85014-837-4
eBook (PDF) ISBN 978-0-85014-839-8
ISSN 2632-0983

If disposing of this product, please recycle the paper responsibly.

We are IntechOpen,
the world's leading publisher of
Open Access books
Built by scientists, for scientists

7,500+
Open access books available

195,000+
International authors and editors

210M+
Downloads

156
Countries delivered to

Our authors are among the
Top 1%
most cited scientists

12.2%
Contributors from top 500 universities

CLARIVATE ANALYTICS
BOOK CITATION INDEX
INDEXED
WEB OF SCIENCE™

Selection of our books indexed in the Book Citation Index
in Web of Science™ Core Collection (BKCI)

Interested in publishing with us?
Contact book.department@intechopen.com

Numbers displayed above are based on latest data collected.
For more information visit www.intechopen.com

# IntechOpen Book Series
# Biochemistry
## Volume 64

## Aims and Scope of the Series

Biochemistry, the study of chemical transformations occurring within living organisms, impacts all of the life sciences, from molecular crystallography and genetics, to ecology, medicine and population biology. Biochemistry studies macromolecules - proteins, nucleic acids, carbohydrates and lipids –their building blocks, structures, functions and interactions. Much of biochemistry is devoted to enzymes, proteins that catalyze chemical reactions, enzyme structures, mechanisms of action and their roles within cells. Biochemistry also studies small signaling molecules, coenzymes, inhibitors, vitamins and hormones, which play roles in the life process. Biochemical experimentation, besides coopting the methods of classical chemistry, e.g., chromatography, adopted new techniques, e.g., X-ray diffraction, electron microscopy, NMR, radioisotopes, and developed sophisticated microbial genetic tools, e.g., auxotroph mutants and their revertants, fermentation, etc. More recently, biochemistry embraced the 'big data' omics systems. Initial biochemical studies have been exclusively analytic: dissecting, purifying and examining individual components of a biological system; in exemplary words of Efraim Racker, (1913 –1991) "Don't waste clean thinking on dirty enzymes." Today, however, biochemistry is becoming more agglomerative and comprehensive, setting out to integrate and describe fully a particular biological system. The 'big data' metabolomics can define the complement of small molecules, e.g., in a soil or biofilm sample; proteomics can distinguish all the proteins comprising e.g., serum; metagenomics can identify all the genes in a complex environment e.g., the bovine rumen.

This Biochemistry Series will address both the current research on biomolecules, and the emerging trends with great promise.

# Meet the Series Editor

Andrei Surguchov, Ph.D., joined Baylor College of Medicine, Houston, TX, as a faculty member in 1992, where he studied the mechanisms of the genetic control of lipid metabolism. At the University of Utah, his research interests focused on cloning new genes encoding retinal proteins. He studied molecular and cellular mechanisms of neurodegenerative diseases and retinal degeneration at Washington University, St. Louis. Currently, his research focuses on the structure-function relationship of proteins involved in neurodegeneration and ocular diseases. Andrei Surguchov is an Editor-in-Chief at Biochemistry Research International and Associate Editor in several biomedical journals.

# Meet the Volume Editor

After completing the University of Avignon doctorate in France, Dr. Muhammad Kamran Khan joined the Department of Food Science at the Government College University Faisalabad in Pakistan. Various research and review articles have been published in peer-reviewed journals during this period. His research focuses on enhancing technical skills and knowledge in food chemistry, with a special focus on polyphenols (bioactive compounds). Dr. Khan has also worked as a Research Fellow at Curtin University in Australia, the University of West Indies, and French Guiana, France. He also won various research grants worth more than 10 million USD. The researcher has also presented his work on different platforms at national and international levels. The professional experience of the researcher greatly helped to edit this book.

# Contents

# Preface

Phytochemistry, the study of phytochemicals, has witnessed rapid and remarkable progress over the past few decades. As scientific inquiry into plant-based compounds expands, researchers have uncovered new and powerful natural products with applications across medicine, agriculture, and environmental sustainability. The current book on *Recent Advances in Phytochemical Research* delves into these cutting-edge developments, offering a comprehensive overview of the latest discoveries and their implications for future applications.

The phytochemicals, particularly bioactive compounds such as alkaloids, flavonoids, terpenoids, and polyphenols, have revealed their diverse roles in human health, from antioxidant and anti-inflammatory properties to anticancer and antimicrobial effects. This book explores how the latest genomics, metabolomics, and bioinformatics methodologies enhance our understanding of plant secondary metabolites and their pharmacological potential. Through these technological innovations, researchers can now identify new bioactive molecules more efficiently and accurately than ever before.

This volume highlights the intersections between traditional knowledge and modern scientific advancements. Many of the discoveries discussed here build upon ancient wisdom, validating the therapeutic potential of plants that have been used for centuries in indigenous and folk medicine. Integrating ethnobotany and modern phytochemistry fosters a more holistic approach to plant research, where cultural heritage and scientific rigor coexist.

The chapters in *Recent Advances in Phytochemical Research* not only present a snapshot of the field's current state but also anticipate future directions. With ongoing advancements in analytical techniques and a deeper understanding of plant genomics, the frontier of phytochemical research is poised for further groundbreaking discoveries. The research featured in this book is a testament to the untapped potential of plants and their powerful contributions to the health and well-being of humanity.

This book is intended for researchers, academics, and students alike, offering an up-to-date review of the field and a source of inspiration for future work in phytochemical sciences. By exploring the many facets of phytochemical research, we aim to inspire continued exploration of the natural world and highlight plant-based science's critical importance in tackling some of the most pressing challenges of our time.

**Dr. Muhammad Kamran Khan**
Department of Food Science,
Government College University Faisalabad,
Faisalabad, Pakistan

Section 1

# Phytochemicals Extraction and Characterization

Chapter 1

# The Biomarker Flavonoid "Rutin" in *Morus* Species: Isolation, Identification, and Characterization

*Nikki Huria, Aparna A. Saraf, Divya Lobo Padinjarekutt, Liviya Gaikwad, Neha Mourya, Dwijalee Deo, Shubham V.U. Tanpathak and Shubham Burande*

## Abstract

Rutin, a quercetin-3-O-rhamnoglucoside, is a naturally occurring flavonol ubiquitous in plants, especially *Morus* species. Rutin, with its antioxidant, antibacterial, and radical scavenging properties, is a promising anti-inflammatory and anticancer agent, potentially improving cardiovascular well-being by reducing inflammation and enhancing blood vessel functionality. Mulberry leaves, rich in nutrients and bioactive components, are used in medicine, human consumption, and animal rearing for their ability to reduce inflammation and act as potent antioxidants. Flavonoids, particularly rutin, possess strong therapeutic and antioxidant properties but have not been fully characterized, necessitating further research to understand their components and pharmacological characteristics. High-performance thin-layer chromatography, high-performance liquid chromatography, nuclear magnetic resonance spectroscopy, mass spectrometry, and crystal isolation can be used to isolate and characterize Rutin, a prevalent flavonoid in *Morus* species. These methods in combination allow for quantitative assessment of rutin content in *Morus* species, uncovering significant variations and highlighting the need for comprehensive phytochemical studies.

**Keywords:** rutin, HPTLC, HPLC, NMR, FTIR, characterization, isolation of rutin crystals, *Morus* species

## 1. Introduction

Flavonoids are an important group of bioactive phytochemicals that are widely acknowledged for their varied biological effects and widespread occurrence in the plant world. The compounds possess a phenolic structure and are ubiquitous in a variety of sources, such as fruits, vegetables, cereals, bark, roots, stems, flowers, tea, and wine. Flavonoids play a crucial role in the pigmentation and pollination processes of plants, contributing to the vibrant colors observed in various fruits and flowers.

In addition to their visual attractiveness, flavonoids serve vital functions in plant defense systems against UV radiation, pests, and diseases, underscoring their importance in ecological interactions and plant adaptation processes [1, 2].

Scientific interest in flavonoids extends beyond their roles in plants, as they exhibit potential health benefits for humans. Research suggests that flavonoids possess antioxidant, anti-inflammatory, and immune-modulating properties, which may contribute to their protective effects against chronic diseases such as cardiovascular disorders and certain cancers. The therapeutic potential of flavonoids has spurred investigations into their dietary intake and bioavailability, aiming to understand how these compounds might promote human health and disease prevention through dietary interventions and pharmaceutical applications [3–5].

Flavonoids are synthesized within plants through the phenylpropanoid pathway, a complex metabolic process involving multiple enzymatic steps and regulatory mechanisms. Flavonoids are synthesized through the conversion of phenylalanine into cinnamic acid, by the action of the enzyme phenylalanine ammonia-lyase, and cinnamic acid undergoes transformations using enzymes such as cinnamate 4-hydroxylase and 4-coumarate CoA ligase. This leads to the formation of chalcone, an intermediary in the synthesis of flavonoids. The enzyme chalcone isomerase converts chalcone into naringenin, which is the precursor for various flavonoids, including flavones, flavonols, flavanones, anthocyanins, and isoflavonoids [6]. The pathway diverges in several directions based on the particular enzymes involved and external variables, which impact the synthesis of distinct kinds and concentrations of flavonoids in various plant tissues and under different circumstances. The knowledge of flavonoid biosynthesis becomes essential for comprehending their functions in plant physiology and adaptations, as well as for implementing their potential advantages in agriculture, nutrition, and medicine [7].

Flavonoids are a diverse group of plant secondary metabolites characterized by their common structure consisting of 15 carbon atoms arranged into two aromatic rings joined by a heterocyclic ring. This basic structure is known as the flavonoid skeleton, which can undergo various modifications such as hydroxylation, methylation, glycosylation, and prenylation. These modifications lead to the extensive structural diversity observed among flavonoids, giving rise to numerous subclasses including flavones, flavonols, flavanones, flavan-3-ols (catechins), anthocyanidins, and isoflavonoids [8].

Flavonoids obtained from plant species of the genus *Morus*, commonly known as mulberry, have been reported for their significant antioxidant capacity. This genus belongs to the family Moraceae. There are around 68 species in the genus *Morus*, with most of them being cultivated in Asia. The significance of mulberry leaf flavonoids lies in their antioxidant, physiological, and pharmaceutical properties. The presence of flavonoids such as rutin, iso-quercitrin, astragalin, kaempferol, and quercetin has been demonstrated to improve animal performance and promote overall health [9]. Metabolic pathways are modulated by these substances, specifically those involved in energy balance. Additionally, they exhibit anabolic properties similar to estrogenic hormones [10]. The administration of mulberry leaf extract in mice resulted in an upregulation of glycolytic enzyme expression and a downregulation of gluconeogenic enzyme expression, thereby effectively modulating glucose balance. The extract derived from mulberry leaves demonstrated a significant reduction in lipid peroxidation and adipocyte size within the liver. The combination of mulberry leaf and fruit extract has been found to effectively reduce inflammation and oxidative stress caused by obesity [11]. The findings indicate that extracts from mulberry leaf and fruit could be used as potential therapeutic treatments for metabolic disorders such as obesity and diabetes. In addition, the extracts' anti-inflammatory and antioxidant properties may lead to enhanced animal health and performance [12].

Rutin, also known as vitamin P, rutoside, quercetin-3-O-rutinoside, and sophorin, is a flavonol glycoside found in *Ruta graveolens*, commonly known as rue. It is found in various plant species, including buckwheat, apples, citrus fruits, and berries, and is known for its antioxidant properties and potential health benefits, including inflammation reduction and cancer prevention [13]. This compound is widely studied for its antioxidant, anti-inflammatory, and vaso-protective properties. Rutin, a commonly found flavonoid in the diet, is a naturally occurring compound that is generally considered to be non-toxic. The displayed attributes of rutin encompass a diverse range of valuable biological properties. These include its capacity to defend against cancer, function as an antioxidant, manage blood sugar levels, minimize inflammation, fight against bacterial and fungal infections, safeguard the brain and nervous system, promote cardiovascular wellness, protect kidneys and liver function, maintain blood health, mitigate symptoms of arthritis, display anthelmintic characteristics, and provide protection to the male reproductive organs. Additionally, this compound has demonstrated potential in enhancing cognitive function and supporting brain health. Furthermore, scientific studies have indicated that it possesses potential anti-aging properties and may play a role in promoting longevity [14].

### 1.1 Structure and biosynthesis

Rutin's chemical structure consists of two main components: a flavonol backbone and a rutinose sugar moiety. The flavonol backbone, quercetin, is a flavonoid compound with a characteristic structure of a diphenyl propane skeleton, hydroxyl groups at positions 3, 5, 7, 3′, 4′, and 5′, and a double bond between carbons 2 and 3 in the C ring. The rutinose sugar, attached *via* a glyosidic bond at position 3 of quercetin, provides solubility and stability to the molecule (**Figure 1**) [16].

Rutin biosynthesis in plants involves the conversion of the precursor compound quercetin to rutin through the addition of a rutinose sugar group catalyzed by specific glycosyltransferases. This process occurs in various plant tissues, particularly in the

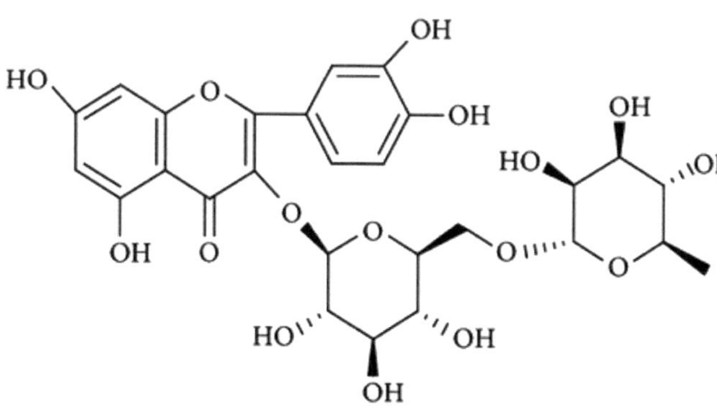


**Figure 1.**
*Chemical structure of rutin [15].*
*Synonyms: Rutoside, Quercetin 3-rutinoside, Vitamin P, etc.*
*Chemical formula: $C_{27}H_{30}O_{16}$;*
*Molecular weight: 610.518;*
*Chemical class: Flavonol.*

leaves, flowers, and seeds, where rutin serves protective roles against UV radiation and oxidative stress [17].

### 1.2 Sources and dietary intake

Rutin is abundantly found in dietary sources such as buckwheat (*Fagopyrum esculentum*), citrus fruits, apples, onions, tea, and some medicinal herbs like *Ginkgo biloba*. The dietary intake of rutin varies depending on cultural dietary habits and food preferences but is estimated to range from a few milligrams to several hundred milligrams per day in typical diets rich in fruits and vegetables.

### 1.3 Pharmacological effects and health benefits

The pharmacological effects of rutin are extensive. Rutin is a significant ingredient found in apples that serves multiple biological functions, thereby supporting the well-known adage "an apple a day keeps the doctor away." The extensive investigation of rutin, a vital phytochemical, is necessary to establish a safe profile for its medicinal use in humans.

The antitumor activity of the combination of rutin and cisplatin against murine ascites Dalton's lymphoma (DL) has been discovered. This combination enhances cytotoxicity and induces apoptosis in DL cells. The described intervention enhances the ability of the host to withstand stress and decreases the levels of glutathione in cancerous cells [18].

A study by Calzada et al. has discovered that rutin derived from *Schinus molle* exhibits anticancer properties. These properties have been observed specifically against the U-937 cell line, which is a type of human leukemic monocyte lymphoma [19].

Patil et al. have determined in their research that rutin and quercetin have the ability to safeguard human lymphocytes from DNA damage caused by radiation. This protective effect is attributed to rutin's antioxidant properties, which enables it to effectively neutralize free radicals and inhibit oxidative stress [20].

A study by Hunyadi et al. reported that rutin, which is the primary constituent of *Morus alba L.* leaf extract, exhibits an antidiabetic effect *in vivo* in type II diabetic rats (**Figure 2**) [22].

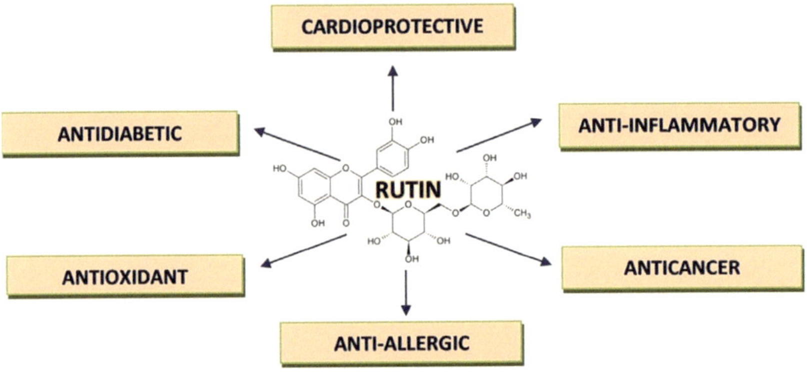


**Figure 2.**
*Health benefits of rutin [21].*

Rutin exhibits a range of pharmacological effects that contribute to its health benefits: [23].

1. Antioxidant activity: Rutin acts as a potent antioxidant, scavenging free radicals and protecting cells from oxidative damage. This property is attributed to the hydroxyl groups on the quercetin backbone, which stabilizes free radicals and inhibits lipid peroxidation.

2. Anti-inflammatory properties: Studies have shown that rutin can suppress inflammatory mediators such as cytokines and prostaglandins, thereby reducing inflammation in various experimental models.

3. Vasoprotective effects: Rutin is known for its beneficial effects on vascular health. It strengthens capillaries and blood vessels, improves microcirculation, and reduces the risk of vascular permeability and edema.

4. Anticancer potential: Research suggests that rutin may have potential anticancer effects, attributed to its antioxidant properties and ability to induce apoptosis in cancer cells.

## 2. Isolation of rutin

The extraction and isolation of phytochemicals, which contribute to biologic activity, is a crucial step in identifying and characterizing plant drugs. Various methods have been used for years, depending on the nature of the phytoconstituent to be isolated and the best isolation technique for the plant drug [24]. Rutin, a phenolic flavonoid, is the rhamno-glucoside of quercetin and precisely follows phenolic compound isolation protocols; after extraction, isolation is typically conducted using chromatographic techniques for further characterization [25]. Different techniques used for the extraction of rutin from various bioresources have been enumerated in the previous studies given below.

One of the main methods that have been used for the isolation entails the exhaustive soxhlation of the powdered/ground plant material with alcohol [26–28]. Thereafter, filtration and evaporation and then subsequent mixing with distilled water (mostly hot) and extraction with petroleum ether and/or chloroform. The aqueous layer is let to stand for up to 72 hours, and the yellow precipitate obtained is separated from solution. After filtering, the precipitate is cleaned using a suitable solution (like chloroform: ethyl acetate: ethanol-50:25:25), and the undissolved portion is dissolved in methanol and filtered; the filtrate is then dried by evaporating to get yellow powder, which is a flavonoid solution containing rutin [29–31]. This flavonoid-rich solution can then be subjected to chromatographic techniques like HPLC [32–34] for further concentration (**Figure 3**).

An alternate method involves taking the impure solution after alcoholic soxhlation to be further refined through modest amounts of cold alcohol extraction, repeated crystallization from aqueous pyridine and/or diluted acetic acid (1%), and/or hot water to give rutin crystals [35].

Another method involves obtaining the hexane extract, and the dried sample is first extracted with hexane at room temperature (25 ± 2°C) for 10 days. The ethanolic extract is then produced by macerating the vegetable material that had been extracted

20 gm of the powdered leaf
↓
Extracted with 80% ethanol
↓
Filtered the extract
↓
Concentrate the extract
↓
Mixed 25ml distilled water
↓
Extracted with pet ether and then chloroform
↓
Aqueous layer collected and stand for 72 hrs at cool place
↓
The precipitate was filtered
↓
Washed with a mixture of chloroform: ethyl acetate: ethanol (50:25:25).
↓
The un-dissolved part of the precipitate was dissolved in hot methanol & filtered
↓
Filtrate was evaporated to dryness to give yellow powder (HS-I).

**Figure 3.**
*Diagrammatic representation of method commonly used for isolation of rutin [30].*

using hexane with ethanol. Additionally, rota-evaporation and filtration are used to concentrate the ethanolic extract, producing a yellow-green material yield. It can then be examined using infrared, 1H, and 13C nuclear magnetic resonance spectroscopy. Rutin (99% purity) can be obtained by recrystallization of this material with methanol [36].

Yet in another alternate method the raw plant material is repeatedly extracted using methanol and 0.1% concentrated hydrochloric acid; then, the filtered extract is concentrated under low pressure and refined by partitioning against ethyl acetate [37].

## 3. Characterization of rutin using various analytical techniques

The characterization of flavonoid structures is closely linked to the elucidation of their spectroscopic spectra, which are obtained using techniques such as nuclear magnetic resonance (NMR), mass spectrometry (MS), spectrophotometric ultraviolet (UV), and infrared (IR). These techniques offer valuable insights into the chemical bonds, functional groups, and overall structure of flavonoids [38].

The characterization of rutin from the leaves of *Morus* species can be conducted using various analytical methods. The identity and purity of isolated rutin were confirmed through the use of HPTLC, HPLC, FTIR, NMR, LCMS, and ultraviolet-visible (UV-vis) spectroscopy techniques. The results obtained from these analyses can offer valuable insights for potential applications of rutin in the pharmaceutical or nutraceutical industries [39].

### 3.1 Characterization of rutin by ultraviolet-visible (UV-vis) spectrophotometry

Ultraviolet-Vis spectrophotometry is a widely recognized technique for identifying and characterizing phenolic compounds, particularly flavonoids. It provides rapid and easily interpretable results, making it advantageous in natural product chemistry.

It expedites the evaluation of bioactive phytochemicals in plant extracts and provides valuable information about the purity and quantity of phenolic chemicals, making it an essential tool for researchers in various scientific disciplines [40].

UV spectroscopy reveals two prominent absorption bands in the presence of flavonoids in methanol spectra. Band I, typically 300–380 nm, is associated with the B-ring cinnamoyl system, while Band II, 240–280 nm, is associated with the A-ring benzoyl system. These bands aid in identifying and quantifying flavonoids in plant extracts. Researchers can compare UV spectra of unknown samples to recognized standards to assess flavonoids' classes and quantities. Flavonoids are classified based on the presence of hydroxyl groups in the A and B rings of the flavonoid structure [41].

The ultraviolet spectrum of rutin in methanolic solution reveals three main absorption bands at 236, 257, and 358.5 nm, indicating a flavonol structure [29]. Savale et al. reported that the rutin solution in methanol exhibited maximum absorption at 257 nm, which was reported as λmax [42]. Hunyadi et al. found that the antidiabetic activity of *Morus alba L.* was attributed to the identification of rutin and chlorogenic acid as biomarkers through UV-vis and HPLC-DAD analysis, which improved glucose tolerance and reduced blood glucose levels in diabetic rats [43].

Pitchaya et al. analyzed the UV spectra of mulberry leaves, revealing rutin as the main flavonoid compound with λmax values at 256 and 353–354. This finding aligns with previous studies, which also identified rutin as a significant flavonoid in mulberry leaves. Rutin is known for its antioxidant and anti-inflammatory properties, suggesting mulberry leaves could be a promising health benefit source [44].

Zhang et al.'s study analyzed rutin's UV spectra to identify its characteristic absorption wavelength for HSCCC detection and HPLC analysis. The chosen wavelength was 257 nm. The results showed rutin had a significant absorption peak at 257 nm, making it suitable for both techniques. This could improve the precision and effectiveness of rutin analysis in diverse samples (**Figure 4**) [45].

Taraba et al. studied UV-visible spectra of rutin in EtOH, revealing two absorption bands at 356 nm (band I) and 258 nm (band II), which is similar to the earlier

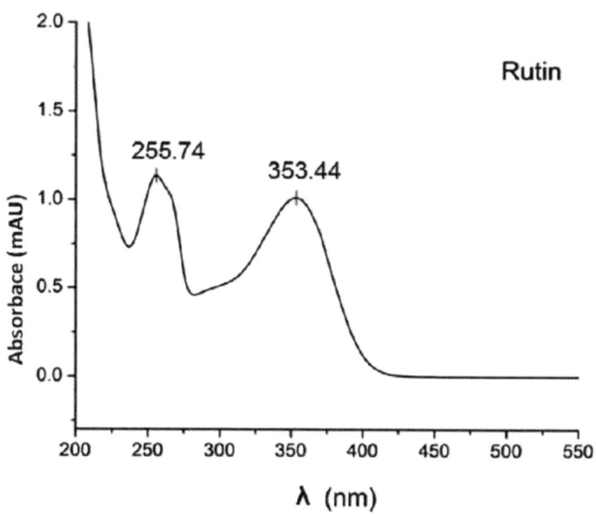


**Figure 4.**
*UV spectra of rutin [4].*

reported studies. Band I correspond to cinnamoyl system between two rings of the flavonoid structure, while band II to the benzoyl moiety. The addition of Triton X-114 (TX114) to the Ru-EtOH combination reduced the intensity of both absorption bands, suggesting interactions between the components [46].

UV absorption is a powerful analytical tool for identifying and characterizing phenolic compounds, particularly flavonoids, in industries such as pharmaceuticals, food and beverage, and cosmetics, providing prompt, interpretable results and quantifying flavonoid concentrations in samples.

### 3.2 Characterization of rutin by Fourier transform infrared spectroscopy (FTIR)

Flavonoids can be identified using chromatographic techniques and mass spectrometry, but Fourier Transform Infrared (FT-IR) spectroscopy is a cost-effective and user-friendly alternative. It eliminates the need for complicated sample preparation and additional reagents [47]. However, interpreting the spectra can be challenging, so a spectrum library or expert consultation may be necessary. These methods can differentiate functional groups, analyze bond structure, and examine conformation, making them valuable for identification and quantification [47]. Understanding the structure of flavonoids is important for determining their antioxidant and antibacterial/antifungal properties, and FT-IR spectroscopy can help analyze their chemical composition and detect substituents that affect functionality [48]. The extracts of medicinal plants are combined with KBr salt using a mortar and pestle, and then compressed into a thin pellet. The pellet is subsequently subjected to infrared spectroscopy for the purpose of identifying the functional groups that are present in the plant extracts. The infrared spectra and peak values are generally recorded using a Fourier transform infrared (FTIR) spectrometer, within the range of 4000 to 400 $cm^{-1}$. The fingerprint region is characterized by prominent absorption bands in the 1500–500 $cm^{-1}$ range, and the FTIR bands within the 4000–1500 $cm^{-1}$ range indicate the presence of a functional group. The observed peaks in the spectra correspond to the vibrations of distinct functional groups, enabling the identification of specific compounds present in the plant extracts.

Fourier transform infrared (FT-IR) spectroscopy is a widely employed analytical technique that enables the examination of structural variations among different samples. However, the interpretation of FT-IR spectra can be challenging when the samples exhibit similar structural characteristics. An example of this is that all spectra include bands that are specific to the stretching vibrations of the C=C ring. However, these bands are found at slightly varying wavenumbers. The vibrational frequency of flavonoids is consistently observed at approximately 1650–1560 $cm^{-1}$. Additionally, all flavonoid groups, except flavanols, exhibit a C–C ring vibration band within the range of 1560–1465 $cm^{-1}$. The spectral peak observed at the wavenumber range of 1475–1400 $cm^{-1}$ is attributed to the presence of the C–C bond in flavonols and anthocyanins compounds. The vibrational (C–C) band of the ring is detected within the range of 1310–1200 $cm^{-1}$ for all flavonoid groups, with the exception of anthocyanins. The presence of the band at 1225–1180 $cm^{-1}$ is exclusive to the spectra of isoflavones, flavonols, and flavanols. The presence of a band at 1046–970 $cm^{-1}$ is exclusively detected in the spectra of isoflavones, flavonols, and anthocyanins. The spectral bands below 970 $cm^{-1}$ display distinct characteristics for each flavonoid group, primarily determined by their ascription rather than their position [49].

The rutin molecule is distinguished by the replacement of the C3 hydroxyl group of quercetin with the disaccharide rutinose (α-L-rhamnopyranosyl-(1 → 6)-β-D-glucopyranose). The primary differences in IR spectra are attributed to the vibration of the C ring and the pyranosyl rings. The presence of a carboxylic acid in rutin can be identified by a distinct band observed between 3550 and 3400 $cm^{-1}$, which corresponds to the O–H bond. This bond is known to contribute to the formation of strong intermolecular hydrogen bonding. The band observed in the range of 1700–1600 $cm^{-1}$ is associated with the stretching vibration of the C=O bond. The presence of a distinct peak in the wavenumber range of 1500–1400 $cm^{-1}$ indicates the presence of C–O and O–H bonds. The rocking vibration frequencies of the C–H bonds in rhamnopyranosyl and glucopyranosyl rings are observed at 1121 and 1168 $cm^{-1}$, respectively. The infrared (IR) spectra exhibit peaks at 1041 and 1093 $cm^{-1}$, indicating the oscillation of the C–O bond between rhamnopyranosyl and glucopyranosyl rings [50]. The infrared (IR) band observed at 999 $cm^{-1}$ can be attributed to the combined wagging vibrations of the C–H bonds in the rhamnopyranosyl ring and the linkage between the rhamno- and glucopyranosyl rings. Additionally, the presence of substitutions in the Ar-H group is indicated by a band observed in the range of 800–600 $cm^{-1}$. The infrared spectrum of rutin offers significant insights into its functional groups and molecular structure.

In their study, Tatke et al. conducted research focused on the isolation of rutin from *Azadirachta indica*. The FTIR analysis revealed the presence of peaks at wavenumbers 3200.29 and 1660.41 $cm^{-1}$, which were identified as the O–H and C=O stretching vibrations, respectively. The results indicate that rutin had been effectively extracted from *Azadirachta indica*. The Fourier transform infrared (FTIR) analysis yielded significant insights into the functional groups that are present in the compound rutin (**Figure 5**) [51, 52].

The spectral peaks of rutin (728, 1062, 1599, 1654, 2878, and 3344 $cm^{-1}$) were analyzed in *Prosopis cineraria* using Fourier-transform infrared spectroscopy. This analysis confirmed the presence of the flavonoid compound rutin in the plant material. The presence of distinct peaks in the data suggested the presence of specific vibrational frequencies that are associated with rutin molecules in *Prosopis cineraria* [53].

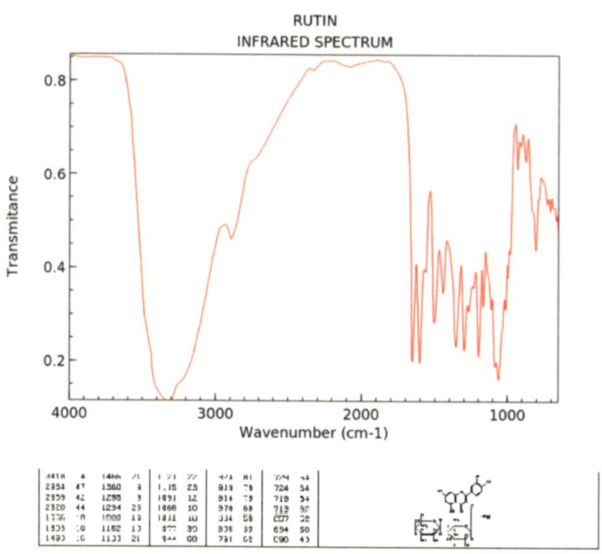


**Figure 5.**
*FTIR spectra of rutin [49].*

The method described can serve as a valuable tool for the identification of rutin in various plant species.

A study on *Terminalia. catappa* by Minsakorn et al. identified quercetin-3-O-rutinoside, a yellow amorphous solid, present in the plant. UV spectra showed a wavelength at 271 nm, while FTIR spectra showed a well-defined peak 3408 $cm^{-1}$ indicating a hydroxyl group attached to the benzene ring. These findings confirmed the presence of rutin in *Terminalia. catappa* [54].

### 3.3 Characterization of rutin by high-performance thin layer chromatography (HPTLC) analysis

High-performance thin layer chromatography (HPTLC) is a valuable analytical tool for evaluating plant quality and formulations containing flavonoids as biomarkers. It helps identify and quantify specific flavonoids like rutin by comparing chromatographic fingerprints from samples and ensuring the quality and standardization of medicinal products. The literature review suggests that a mobile phase of ethyl acetate, formic acid, glacial acetic acid, and water in different ratios is commonly used for high-performance thin-layer chromatography (HPTLC) analysis of rutin, especially in *Morus* species. This mobile phase has been reported to demonstrate effective resolution and separation of rutin from other compounds, but it is important to consider that the mobile phase composition may require adjustment depending on the specific characteristics of the samples under analysis [55].

Based on the findings of Sánchez-Salcedo et al. as well as Polumackanycz et al., rutin, a biomarker found in the leaves of various *Morus* species, can be used to authenticate the medicinal properties of the mulberry plant. The research conducted by these two groups emphasizes the significance of rutin as a dependable biomarker for differentiating between various species of mulberry [56, 57].

A study by Jan et al. reported the presence of rutin and chlorogenic acid in *Morus alba L.* tea infusions using the HPTLC method, using ethyl acetate/formic acid/acetic acid/water (100:11:11:26, *v/v/v/v*) as the mobile phase. The study found that rutin had a higher concentration than chlorogenic acid, and is well known for its antioxidant properties and potential health benefits. Consistent consumption of this tea may lead to health benefits due to the presence of these compounds (**Figure 6**) [59].

The study by Kathia et al. evaluated the phenolic content, rutin, and antioxidant and antibacterial activity of methanolic extracts from *Verbesina sphaerocephala* leaves and flowers. HPTLC was used to detect the presence of flavonoid rutin in all extracts, and the mobile phase used was a mixture of ethyl acetate, formic acid, acetic acid, and water, revealing elevated levels of phenolic and flavonoid compounds showing high antioxidant activity [60].

The high-performance thin-layer chromatography (HPTLC) technique is widely recognized as a valuable method for the qualitative analysis of flavonoids in plant extracts. This method is highly sensitive and specific, allowing for the detection and precise identification and quantification of flavonoids, thereby offering valuable data for subsequent research on the potential health advantages associated with these plant extracts [61].

Mulberry leaves are a rich source of flavonoids, providing antioxidant, anti-tumor, anticancer, and anti-inflammatory benefits. HPTLC analysis can assess the quality of *Morus* species by quantifying rutin in unprocessed herbal remedies. This method ensures the potency and efficacy of mulberry leaf supplements, as rutin is a significant flavonoid with numerous health benefits. Regular monitoring of rutin levels can

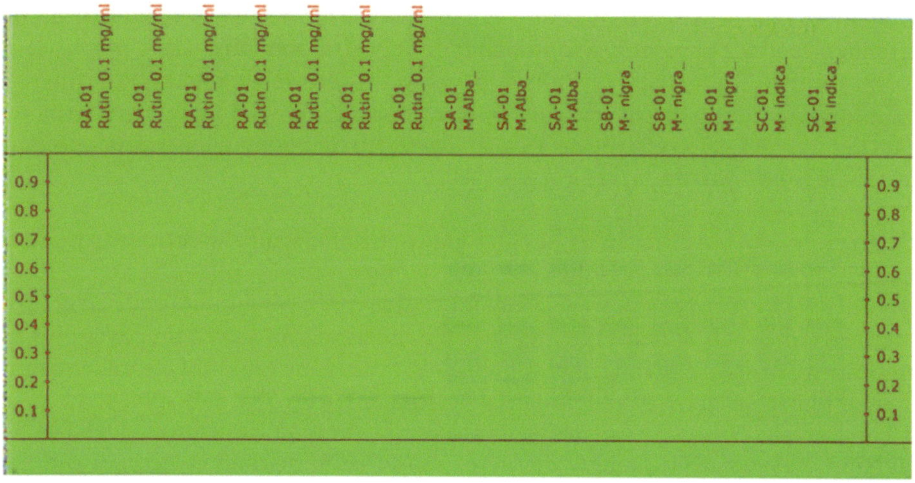


**Figure 6.**
*HPTLC chromatogram of rutin in* Morus *species [58].*

detect potential adulteration or contamination, thereby contributing to the quality control and efficacy of herbal products.

### 3.4 Characterization of rutin by high-performance liquid chromatography (HPLC) analysis

The high-performance liquid chromatography (HPLC) method is a preferred chromatographic technique for analyzing flavonoids due to its efficient time-saving nature, elimination of derivatization, and its ability to be conducted at room temperature, ensuring the preservation of delicate compounds such as flavonoids without the risk of decomposition at higher temperatures [62]. It involves isolating phenolic chemicals from the samples and using HPLC with a gradient mobile phase. Target flavonoids are identified and quantified by comparing them to reference standards using a polar organic solvent and mild acid [63].

Researchers have widely studied flavonoid components in mulberry species, detecting phenolic acids and flavonol glycosides in leaves using HPLC analysis. Thabti et al. identified compounds such as rutin, kaempferol-7-O-glucoside, and quercetin-3-O-rhamnoside-7-O-glucoside in mulberry leaves. Furthermore, it was determined that the leaves of *Myrionecta rubra* exhibited the highest overall flavonoid content when compared to the other two mulberry species [64].

A study by Ju et al. used high-performance liquid chromatography (HPLC) to analyze the flavonoid content in Korean mulberry leaves. The results showed that the leaves had high levels of quercetin and rutin, known for their antioxidant properties. These flavonoids, which are abundant in Korean mulberry leaves, could potentially contribute to the health benefits associated with flavonoid-rich foods. The findings suggest that these compounds could significantly influence the consumption of these foods [65].

A study by Sarita et al. reported that rutin and quercetin were the most abundant flavonoids in mulberry leaf extract, with flavonoids such as quercetin, isoquercetin, and luteolin also present. The RP-HPLC methodology was used to identify these components, using HPLC with a UV-visible detector and a C18 column for chromatographic separation. Isocratic elution was performed using mobile phase of acetonitrile and 0.1 percent v/v solution of glacial acetic acid [66].

HPLC is a vital analytical technique in the pharmaceutical industry, ensuring accurate data for drug discovery, development, formulation, and quality control. A validated method helps pharmaceutical analysts deliver better-quality products and meet required standards.

### 3.5 Characterization of rutin by nuclear magnetic resonance spectroscopy (NMR)

NMR spectroscopy is a powerful tool for analyzing flavonoid molecule structures in complex materials. It helps identify specific chemical groups within flavonoids and determine their spatial arrangement, providing crucial information for understanding their biological activity and potential health benefits [67].

The study by Selvaraj et al. analyzed flavonoid compounds using TLC, PTLC, and HPLC. UV-vis, FTIR, and NMR were used to examine their structures and chemical interactions. Rutin and quercetin were reported, showing significant antioxidant radical scavenging action. Nuclear magnetic resonance (1H and 13C) spectra were acquired in CDCl3 using a Bruker NMR Avance and TCI cyroprobe at 600 MHz [68].

Ghareeb et al. conducted a study to examine several extracts of *Morus alba* leaves and assess their antioxidant properties. Rutin and other flavonoids were detected using UV-vis, FTIR, and NMR spectroscopic investigations. The research discovered that the ethyl acetate extract had the greatest level of antioxidant activity, possibly attributed to the presence of rutin [69].

The study by Shuang et al. explored the antioxidant properties of *Morus alba* leaves and their mechanism of scavenging free radicals. Rutin was extracted by liquid-liquid extraction, isolated by high-speed counter-current chromatography, and identified by NMR. Rutin was identified as the primary antioxidant in mulberry leaves with strong efficacy in combating DPPH·and $ABTS^+$ radicals [70].

Fatima et al. conducted a study on the characterization and quantitation of flavonoids and phenolic components extracted from lyophilized mulberry leaves from two different cultivars. They identified the major flavonoid, rutin, using a hybrid IT-TOF MS system coupled to HPLC, and performed NMR analysis [71].

NMR spectroscopy is a valuable tool for understanding the structure of complex molecules, particularly flavonoids such as rutin and other phenolic components found in mulberry leaves.

### 3.6 Characterization of rutin by liquid chromatography-mass spectroscopy (LC: MS)

LC-MS is a widely used technique for identifying flavonoids and flavonoid glycosides, combining chromatography's separation capabilities with precise detection of compound-specific precursors and fragments. It is particularly useful for examining complex plant extract combinations, where conventional techniques may lack sensitivity to measure low flavonoids levels, making it a valuable tool in phytochemical analysis and drug discovery [72].

HPLC can be integrated using multiple detectors like MS/MS and is a widely used method for identifying species and ensuring the quality of herbal medicines, specifically applied to ginkgo and mulberry leaves, including *Morus nigra L. and Morus alba L.* [73, 74].

The study by Xiaoyun et al. developed a precise and cost-effective chromatographic fingerprinting profile for evaluating mulberry leaves' quality using HPLC, DAD, and MS methods. The study used an Agilent Technologies 1200 analytical HPLC system with a quaternary pump autosampler and triple quadrupole mass spectrometer for elution at 30°C using a Poroshell 120 EC-C18 column, with a mobile phase of 0.1% formic acid and acetonitrile. The method was validated and effectively used to identify rutin in mulberry leaves [75].

Rutin was reported in the leaves of two mulberry cultivars, Yai-Burirum (YB) and Khunphai (KP), which were processed into green tea (GT) and black tea (BT). The LCMS analysis showed that the rutin content in GT and BT extracts was significantly higher than the FL extract. The highest rutin content was found in the YB GT extract [76].

Research conducted by Ling Wan et al. using LCMS has shown that mulberry leaf ethanolic extract includes many flavonoids, including rutin, isoquercetin, and astragalin, that have the potential to inhibit xanthine oxidase enzyme, leading to extract's potent antioxidant and anti-inflammatory capabilities [77].

A study by Yu Meng et al. found that mulberry leaves contain numerous polyphenols, including protocatechuic acid, chlorogenic acid, nicotiflorin, rutin, and astragalin. The most abundant bioactive flavonoid in mulberry leaf polyphenol extract (MLPE) was rutin, which regulates autophagy in HepG2 cells by influencing p53 signaling and also has antioxidative and anticancer properties [78].

Wang et al. performed a research on mulberry leaves using qualitative LC-ESI-QTOF analysis, which revealed a strong correlation between mulberry leaves' antioxidant activities and the levels of isoquercitrin, chlorogenic acid, and rutin in free phenolic extracts. Additionally, syringic acid and rutin were revealed to be the primary contributors to the antioxidant activities of bound phenolic fractions [79].

Liquid chromatography-mass spectrometry (LC-MS) could be a useful analytical technique for quantifying rutin concentration in mulberry leaf extracts, with further research exploring their potential physiological benefits.

## 4. Conclusion

Rutin, a flavonoid-derived compound from quercetin glycosides, has various biological activities such as antioxidant, anti-radical, estrogenic, anti-inflammatory, antiviral, anticancer, cytotoxic, and organ-protective effects. This chapter highlights the importance of natural product chemistry in identifying bioactive compounds like rutin with potential health benefits, emphasizing the significance of these methods in the field of natural product chemistry. Rutin can be identified and characterized using techniques like UV-VIS spectrophotometry, FTIR, HPTLC, HPLC, NMR, and LCMS, providing valuable data for future research in the pharmaceutical and nutraceutical sectors.

The identification, isolation, and characterization of flavonoid rutin from *Morus* species have been achieved using various techniques and methodologies. HPLC is commonly used for extraction, especially from plant species. Spectroscopic techniques such as UV-Vis, FTIR, and NMR are used for structural elucidation. Rutin, a flavonoid compound identified in *Morus* leaves, has potential therapeutic properties such as antioxidant and anti-inflammatory effects. Further research is needed to understand its pharmacological activities and potential applications in medicine, particularly in cancer research, which could lead to innovative therapies for various diseases.

## Conflict of interest

The authors declare no conflict of interest.

## Author details

Nikki Huria*, Aparna A. Saraf, Divya Lobo Padinjarekutt, Liviya Gaikwad, Neha Mourya, Dwijalee Deo, Shubham V.U. Tanpathak and Shubham Burande
The Institute of Science, Dr. Homi Bhabha State University, Mumbai, India

*Address all correspondence to: resbot20.nikki@iscm.ac.in

IntechOpen

© 2024 The Author(s). Licensee IntechOpen. This chapter is distributed under the terms of the Creative Commons Attribution License (http://creativecommons.org/licenses/by/4.0), which permits unrestricted use, distribution, and reproduction in any medium, provided the original work is properly cited.

## References

[1] Rodríguez-García C, Sánchez-Quesada C, Gaforio JJ. Dietary flavonoids as cancer chemopreventive agents: An updated review of human studies. Antioxidants. 2019;**8**(5):137. DOI: 10.3390/antiox8050137

[2] Nabavi SM, Šamec D, Tomczyk M, Milella L, Russo D, Habtemariam S, et al. Flavonoid biosynthetic pathways in plants: Versatile targets for metabolic engineering. Biotechnology Advances. 2018;**38**:107216. DOI: 10.1016/j.biotechadv.2018.11.005

[3] Arora S, Itankar P. Extraction, isolation and identification of flavonoid from chenopodium album aerial parts. Journal of Traditional and Complementary Medicine. 2018;**8**(4):476-482. DOI: 10.1016/j.jtcme.2017.10.002

[4] Rodríguez De Luna SL, Ramírez-Garza RE, Serna Saldívar SO. Environmentally friendly methods for flavonoid extraction from plant material: Impact of their operating conditions on yield and antioxidant properties. The Scientific World Journal. 2020;**2020**(1):6792069. DOI: 10.1155/2020/6792069

[5] Rodríguez De Luna SL, Ramírez-Garza RE, Serna Saldívar SO. The immunomodulatory and anti-inflammatory role of polyphenols. Nutrients. 2018;**10**:E1618. DOI: 10.3390/nu10111618

[6] Nabavi SM, Šamec D, Tomczyk M, Milella L, Russo D, Habtemariam S, et al. Flavonoid biosynthetic pathways in plants: Versatile targets for metabolic engineering. Biotechnology Advances. 2020;**38**:107316. DOI: 10.1016/j.biotechadv.2018.11.005

[7] Stafford HA. Possible multienzyme complexes regulating the formation of C6-C3 phenolic compounds and lignins in higher plants. In: Recent Advances in Phytochemistry. Vol. 8. Amsterdam, Netherlands: Elsevier; 1974. pp. 53-79. DOI: 10.1016/B978-0-12-612408-8.50009-3

[8] Dias MC, Pinto DCGA, Silva AMS. Plant flavonoids: Chemical characteristics and biological activity. Molecules (Basel, Switzerland). 2021;**26**(17):5377. DOI: 10.3390/molecules26175377

[9] Yu Y, Li H, Zhang B, Wang J, Shi X, Huang J. Nutritional and functional components of mulberry leaves from different varieties: Evaluation of their potential as food materials. International Journal of Food Properties. 2018;**21**(2018):1495-1507. DOI: 10.1080/10942912.2018.148983

[10] Stoldt A, Derno M, Das G, Weitzel JM, Wolffram S, Metges CC. Effects of rutin and buckwheat seeds on energy metabolism and methane production in dairy cows. Journal of Dairy Science. 2016;**99**(2016):2161-2168. DOI: 10.3168/jds.2015-10143

[11] Berger LM, Blank R, Zorn F, Wein S, Metges CC, Wolffram S. Ruminal degradation of quercetin and its influence on fermentation in ruminants. Journal of Dairy Science. 2015;**98**:5688-5698. DOI: 10.3168/jds.2015-9633

[12] Lim HH, Lee SO, Kim SY, Yang S, Lim Y. Anti-inflammatory and antiobesity effects of mulberry leaf and fruit extract on high fat diet-induced obesity. Experimental Biology and Medicine. 2013;**238**(2013):1160-1169. DOI: 10.1177/1535370213498982

[13] Gawlik-Dziki U. Changes in the antioxidant activities of vegetables as a consequence of interactions between active compounds. Journal of Functional Foods. 2012;**4**(4):872-882

[14] Prasad R, Prasad S. A review on the chemistry and biological properties of rutin, a promising nutraceutical agent. Asian Journal of Pharmacy and Pharmacology. 2019;**5**:1-20. DOI: 10.31024/ajpp.2019.5.s1.1

[15] Tian SG, Xin LD, Upur H. High-performance thin-layer chromatographic quantification of rosmarinic acid and rutin in abnormal savda munziq. Journal of Chemistry. 2013;**2013**(1):898167

[16] Harborne JB. Phytochemical Methods, a Guide to Modern Techniques of Plant Analysis. New Delhi, India: Springer Pvt. Ltd; 1998. DOI: 10.1046/j.1365-3059.1999.00318.x

[17] Ganeshpurkar A, Saluja AK. The pharmacological potential of rutin. Saudi Pharmaceutical Journal. 2017;**25**(2):149-164. DOI: 10.1016/j.jsps.2016.04.025

[18] Prasad R, Prasad SB. Antitumor activity of rutin-cisplatin in combination and its protective effect against hematotoxicity. Research Journal of Life Sciences, Bioinformatics, Pharmaceutical and Chemical Sciences. 2018;**4**:42-56. DOI: 10.26479/2018.0406.04

[19] Calzada F, Solares-Pascasio JI, Valdes M, Garcia-Hernandez N, Velázquez C, Ordoñez-Razo RM, et al. Antilymphoma potential of the ethanol extract and rutin obtained of the leaves from Linn. *Schinus molle*. Pharmacognosy Research. 2018;**10**:119-123. DOI: 10.4103/pr.pr_90_17

[20] Patil SL, Swaroop K, Kakde N, Somashekarappa HM. In vitro protective effect of rutin and quercetin against radiation induced genetic damage in human lymphocytes. Indian Journal of Nuclear Medicine. 2017;**32**:289-295. DOI: 10.4103/ijnm.IJNM_30_17

[21] Semwal R, Joshi SK, Semwal RB, Semwal DK. Health benefits and limitations of rutin - A natural flavonoid with high nutraceutical value. Phytochemistry Letters. 1 Dec 2021;**46**:119-128. DOI: 10.1016/j.phytol.2021.10.006

[22] Hunyadi A, Martins A, Hsieh T-J, Seres A, Zupko I. Chlorogenic acid and rutin play a major role in the *in vivo* anti-diabetic activity of leaf extract on type II *Morus albadiabetic* rats. PLoS One. 2012;7:1-6. DOI: 10.1371/journal.pone.0050619

[23] Patel K, Patel DK. The beneficial role of rutin, a naturally occurring flavonoid in health promotion and disease prevention: A systematic review and update. In: Bioactive Food as Dietary Interventions for Arthritis and Related Inflammatory Diseases. Arizona: Academic Press; 2019. pp. 457-479. DOI: 10.1016/B978-0-12-813820-5.00026-X

[24] Sindhu KR, Kaur P, Manshu S, Kaur P, Goyal A, Bala R, et al. Phytochemicals: Extraction, isolation methods, identification and therapeutic uses: A review. Plant Archives. 2021;**21**(Suppliment-1):174-184. DOI: 10.51470/plant

[25] Hoda M, Hemaiswarya S, Doble M. Phenolic phytochemicals: Sources, biosynthesis, extraction, and their isolation. In: Role of Phenolic Phytochemicals in Diabetes Management. Singapore: Springer; 2019. pp. 13-44. DOI: 10.1007/978-981-13-8997-9_2

[26] Awad ZJ, Naser EH. Isolation and identification of bioactive flavonoids

(genistein, rutin) from *Portulaca oleracea* L. cultivated in Iraq. Kerbala Journal of Pharmaceutical Sciences. 2015;**10**

[27] Ogunkunle ATJ, Oyelakin TM, Enitan AO, Oyewole FE. A quantitative documentation of the composition of two powdered herbal formulations (antimalarial and haematinic) using ethnomedicinal information from Ogbomoso, Nigeria. Evidence-Based Complementary and Alternative Medicine. 2014;**2014**:751291. DOI: 10.1155/2014/751291

[28] Srinivasan R, Natarajan D, Shivakumar MS. Antioxidant compound quercetin-3-O-α-L-rhamnoside (1→6)-β-D-glucose (Rutin) isolated from ethyl acetate leaf extracts of Memecylon edule Roxb (Melastamataceae). Free Radicals and Antioxidants. 2015;**5**(1):35-42. DOI: 10.5530/fra.2015.1.6

[29] Soni H, Bhopal D, Pradesh M, Malik IJ, Malik J. Isolation, characterization and radical scavenging activities of rutin isolated from leaves of *Annona squamosa*. Journal of Pharmacognosy and Phytochemistry. 2022;**11**(2):222-228. DOI: 10.22271/phyto

[30] Soni H, Malik J, Yadav AP, Yadav B. Characterization of rutin isolated by leaves of *Annona squamosa* by modern analytical techniques. European Journal of Biomedical and Pharmaceutical Sciences. 2018;**5**(6):484-489. Available from: www.ejbps.com

[31] Al-Mahdawe MM, Al-Mallah MK, Ahmad TA. Isolation and identification of rutin from tissues cultures of *Ruta graveolens* L. Journal of Pharmaceutical Sciences and Research. 2018;**10**(6):1517-1520

[32] Abdel-Naim AB, Alghamdi AA, Algandaby MM, Al-Abbasi FA, Al-Abd AM, Eid BG, et al. Rutin isolated from *Chrozophora tinctoria* enhances bone cell proliferation and ossification markers. Oxidative Medicine and Cellular Longevity. 2018;**2018**:5106469. DOI: 10.1155/2018/5106469

[33] Al-Majmaie SHA. Phytochemical and Antimicrobial Studies on Selected Medicinal Plants from the Iraqi Flora: Citrus grandis, *Citrus sinensis* and *Ruta chalepensis*. United Kingdom: Liverpool John Moores University; 2019

[34] Choi SJ, Lee SN, Kim K, Joo DH, Shin S, Lee J, et al. Biological effects of rutin on skin aging. International Journal of Molecular Medicine. 2016;**38**(1):357-363

[35] Sando CE, Lloyd JU. The isolation and identification of rutin from the flowers of elder (*Sambucus Canadensis* L.). Journal of Biological Chemistry. 1924;**58**(3):737-745. DOI: 10.1016/s0021-9258(18)85333-9

[36] Machado DG, Bettio LEB, Cunha MP, Santos ARS, Pizzolatti MG, Brighente IMC, et al. Antidepressant-like effect of rutin isolated from the ethanolic extract from *Schinus molle* L. in mice: Evidence for the involvement of the serotonergic and noradrenergic systems. European Journal of Pharmacology. 2008;**587**(1-3):163-168. DOI: 10.1016/j.ejphar.2008.03.021

[37] Andemen M, Opheim S, Aksnes DW, Frgystein NA. Structure of petanin, an acylated anthocyanin isolated from *Solanum tuberosum*, using homo-and hetero-nuclear two-dimensional nuclear magnetic resonance techniques. Phytochemical Analysis. 1991;**2**:230-236

[38] Awouafack MD, Tane P, Morita H. Isolation and Structure Characterization of Flavonoids. London, UK: InTech; 2017. DOI: 10.5772/67881

[39] Soni H, Malik J. Isolation, characterization and radical scavenging activities of rutin isolated from leaves of *Annona squamosa*. Journal of Pharmacognosy and Phytochemistry. 2022;**11**(2):222-228

[40] Sattiraju VK, Jitta SR, Pratap R, Pai V, Chandrashekar KS. Characterization of flavonoids isolated from *Trigonellafoenum-Graecum* using UV shift reagent. Rasayan Journal of Chemistry. 2021;**14**(3):1845-1850. DOI: 10.31788/RJC.2021.1436334

[41] Picollo M, Aceto M, Vitorino T. UV-vis spectroscopy. Physical Sciences Reviews. 2019;**4**:20180008. DOI: 10.1515/psr-2018-0008

[42] Savale S. UV spectrophotometric method development and validation for quantitative estimation of rutin. Hygeia Journal for Drugs and Medicines. 2018;**9**(2):7-10. DOI: 10.15254/H.J.D.Med.9.2018.168

[43] Hunyadi A, Martins A, Hsieh T-J, Seres A, Zupko I. Chlorogenic acid and rutin play a major role in the *in vivo* anti-diabetic activity of *Morus alba* leaf extract on type II diabetic rats. PLoS One. 2012;7(11):e50619. DOI: 10.1371/journal.pone.0050619

[44] Pothinuch P, Tongchitpakdee S. Phenolic analysis for classification of mulberry (*Morus* spp.) leaves according to cultivar and leaf age. Journal of Food Quality. 2019;**2019**(1):1-11. DOI: 10.1155/2019/2807690

[45] Zhang P, Zhu K-L, Zhang J, Li Y, Zhang H, Wang Y. Purification of flavonoids from mulberry leaves via high-speed counter-current chromatography. PRO. 2019;7(2):91. DOI: 10.3390/pr7020091

[46] Nie K, Zhonghai Tang XW, Xiaona X, Liang Y, Li H, Rao L. Optimization of total flavonoids extraction from mulberry leaf using an ethanol-based solvent system. Journal of Medicinal Plants Research. 2012;**6**(12):2373-2380. DOI: 10.5897/JMPR11.993

[47] Geraldes CFGC. Introduction to infrared and Raman-based biomedical molecular imaging and comparison with other modalities. Molecules. 2020;**25**(23):5547. DOI: 10.3390/molecules25235547

[48] Gniadecka M, Wulf HC, Nielsen OF, Christensen DH, Hercogova J. Distinctive molecular abnormalities in benign and malignant skin lesions: Studies by Raman spectroscopy. Photochemistry and Photobiology. 1997;**66**(4):418-423. DOI: 10.1111/j.1751-1097.1997.tb03167.x

[49] Krysa M, Szymańska-Chargot M, Zdunek A. FT-IR and FT-Raman fingerprints of flavonoids – A review. Food Chemistry. 1 Nov 2022;**393**:133430. DOI: 10.1016/j.foodchem.2022.133430

[50] Paczkowska M, Lewandowska K, Bednarski W, Mizera M, Podborska A, Krause A, et al. Application of spectroscopic methods for identification (FT-IR, Raman spectroscopy) and determination (UV, EPR) of quercetin-3-O-rutinoside. Experimental and DFT based approach. Spectrochimica Acta - Part A: Molecular and Biomolecular Spectroscopy. 2015;**140**:132-139. DOI: 10.1016/j.saa.2014.12.050

[51] Tatke P, Desai S, Gabhe SY. Isolation of quercetin-3-O-β-D-glucoside from *Azadirachta indica*. American Journal of Phytomedicine and Clinical Therapeutics. 2014;**2**(7):870-876

[52] Auha NA, Alauhdin M. Development and validation of infrared spectroscopy methods for rutin compound analysis. Indonesian Journal of Chemical Science. 2021;**10**(2):102-107

[53] Bishnoi RS, Kumar M, Shukla AK, Jain CP. Comparative fingerprint and extraction yield of *Prosopis cineraria* (Lin.) Druce. leaves with phenolic compounds (gallic acid) & flavonoids (rutin). Journal of Drug Delivery and Therapeutics. 2019;**9**(3-s):560-568

[54] Minsakorn S, Watthanadirek A, Poolsawat N, Puttarak P, Chawengkirttikul R, Anuracpreeda P. The anthelmintic potentials of medicinal plant extracts and an isolated compound (rutin, C27H30O16) from *Terminalia catappa* L. against Gastrothylax crumenifer. Veterinary Parasitology. 2021;**291**:109385. DOI: 10.1016/j.vetpar.2021.109385

[55] Dighe VV, Sane RT, Menon S, Gokarn VG, Gursale A. A high performance thin layer chromatographic method for quantitative determination of rutin in leaf powder of *Morus alba* Linn. Indian Drugs-Bombay. 2007;**44**(2):117

[56] Sanchez-Salcedo EM, Tassotti M, Del Rio D, Hernandez F, Martinez JJ, Mena P. (poly) phenolic compounds and antioxidant activity of white (*Morus alba*) and black (*Morus nigra*) mulberry leaves: Their potential for new products rich in phytochemicals. Journal of Functional Foods. 2015;**18**:1039-1046. DOI: 10.1016/j.jff.2015.03.053

[57] Polumackanycz M, Sledzinski T, Goyke E, Wesolowski M, Viapiana A. A comparative study on the phenolic composition and biological activities of *Morus alba* L. commercial samples. Molecules. 2019;**24**(17):3082. DOI: 10.3390/molecules24173082

[58] Huria N, Saraf A, Saraf A. Estimation of rutin from *Morus* species using validated HPTLC method. International Journal of Pharmaceutical Sciences. 2024;**2**(3):565-577. DOI: 10.5281/zenodo.10954737

[59] Jan B, Zahiruddin S, Basist P, Khan MU, Abass S, Dewangan RP, et al. Separation and identification of antioxidant and antihyperglycemic constituents from *Morus alba* infusion by TLC-MS-bioautography and UPLC-MS. Journal of Food and Nutrition Research. 2021;**9**(12):670-690. DOI: 10.12691/jfnr-9-12-8

[60] Rodríguez-Valdovinos KY, Salgado-Garciglia R, Vázquez-Sánchez M, Álvarez-Bernal D, Oregel-Zamudio E, Ceja-Torres LF, et al. Quantitative analysis of rutin by HPTLC and *in vitro* antioxidant and antibacterial activities of phenolic-rich extracts from Verbesina sphaerocephala. Plants. 2021;**10**(3):475. DOI: 10.3390/plants1003

[61] Rodríguez-Valdovinos KY, Salgado-Garciglia R, Vázquez-Sánchez M, Álvarez-Bernal D, Oregel-Zamudio E, Ceja-Torres LF, et al. Quantitative analysis of rutin by HPTLC and *in vitro* antioxidant and antibacterial activities of phenolic-rich extracts from Verbesina sphaerocephala. Plants. 2021;**10**(3):475. DOI: 10.3390/plants10030475

[62] Numonov SR, Qureshi MN, Aisa HA. Development of HPLC protocol and simultaneous quantification of four free flavonoids from Dracocephalum heterophyllum benth. International Journal of Analytical Chemistry. 2015;**2015**:503139. DOI: 10.1155/2015/503139

[63] Mizzi L, Chatzitzika C, Gatt R, Valdramidis V. HPLC analysis of phenolic compounds and flavonoids with overlapping peaks. Food Technology and Biotechnology. 2020;**58**(1):12-19. DOI: 10.17113/ftb.58.01.20.6395

[64] Thabti I, Elfalleh W, Tlili N, Ziadi M, Campos MG, Ferchichi A. Phenols, flavonoids, and antioxidant and antibacterial activity of leaves

and stem bark of *Morus* species. International Journal of Food Properties. 2014;**17**(4):842-854

[65] Ju WT, Kwon OC, Kim HB, et al. Qualitative and quantitative analysis of flavonoids from 12 species of Korean mulberry leaves. Journal of Food Science and Technology. 2018;**55**:1789-1796. DOI: 10.1007/s13197-018-3093-2

[66] Garg S, Chahal R, Kaushik D, Kumar R, Mittal V. RP-HPLC method development and validation for simultaneous estimation of rutin and quercetin in *Morus alba* L. leaf extract. Research Journal of Pharmacy and Technology. 2023;**16**(5):232

[67] Zou L, Li H, Ding X, Liu Z, He D, Kowah JAH, et al. A review of the application of spectroscopy to flavonoids from medicine and food homology materials. Molecules. 2022;**27**:7766. DOI: 10.3390/molecules27227766

[68] Selvaraj K, Chowdhury R, Bhattacharjee C. Isolation and structural elucidation of flavonoids from aquatic Fern *Azolla Microphylla* and evaluation of free radical scavenging activity. International Journal of Pharmacy and Pharmaceutical Sciences. 2013;**5**(3):743-749

[69] Ghareeb MA, Ahmed WS, Refahy LA, Abdou AM, Hamed MM, Abdel-Aziz MS. Isolation and characterization of the bioactive phenolic compounds from *Morus Alba* leaves growing in Egypt. Pharmacology. 2016;**3**:157-167

[70] Guo S, Dongyu G, Liu C, Tang S, Wang Y, Yang Y. High-speed counter-current chromatographic separation and thermodynamic mechanism of an antioxidant from *Morus alba* leaves. Biochemical Systematics and Ecology. 2023;**107**:104617. DOI: 10.1016/j.bse.2023.104617

[71] Fatima M, Dar MA, Dhanavade MJ, Abbas SZ, Bukhari MN, Arsalan A, et al. Biosynthesis and pharmacological activities of the bioactive compounds of white mulberry (*Morus alba*): Current paradigms and future challenges. Biology. 2024;**13**(7):506. DOI: 10.3390/biology13070506

[72] Soares JC, Rosalen PL, Lazarini JG, Massarioli AP, da Silva CF, Nani BD, et al. Food Chemistry. 2019;**281**:178-188

[73] Sanchez-Salcedo EM, Tassotti M, Del Rio D, Hernandez F, Martinez JJ, Mena P. (poly)phenolic fingerprint and chemometric analysis of white (*Morus alba* L.) and black (*Morus nigra* L.) mulberry leaves by using a non-targeted UHPLC-MS app approach. Food Chemistry. 2016;**212**:250-255

[74] Yang J, Wen H, Zhang L, Zhang X, Fu Z, Li J. The influence of ripening stage and region on the chemical compounds in mulberry fruits (*Morus atropurpurea* Roxb.) based on UPLC-QTOF-MS. Food Research International. 2017;**100**(Pt 2):159-165

[75] He X, Chen X, Ou X, Ma L, Xu W, Huang K. Evaluation of flavonoid and polyphenol constituents in mulberry leaves using HPLC fingerprint analysis. International Journal of Food Science & Technology. 2020;**55**(2):526-533

[76] Panyatip P, Padumanonda T, Yongram C, Kasikorn T, Sungthong B, Puthongking P. Impact of tea processing on tryptophan, melatonin, phenolic and flavonoid contents in mulberry (*Morus alba* L.) leaves: quantitative analysis by LC-MS/MS. Molecules. 2022;**27**(15):4979. DOI: 10.3390/molecules27154979

[77] Wan L, Chen G, Jian S, Yin HJ, Zhu H. Antioxidant and xanthine oxidase inhibitory properties and LC-MS/MS identification of compounds of ethanolic extract from mulberry leaves. Acta Scientiarum Polonorum. Technologia Alimentaria. 2018;**17**(4):313-319. DOI: 10.17306/J.AFS.2018.0587

[78] Yu M-H, Tsai M-C, Wang C-C, Wu S-W, Chang Y-J, Wu C-H, et al. Mulberry leaf polyphenol extract and rutin induces autophagy regulated by p53 in human hepatoma HepG2 cells. Pharmaceuticals. 2021;**14**(12):1310. DOI: 10.3390/ph14121310

[79] Wang Z, Tang C, Xiao G, et al. Comparison of free and bound phenolic compositions and antioxidant activities of leaves from different mulberry varieties. BMC Chemistry. 2021;**15**:21. DOI: 10.1186/s13065-021-00747-0

Chapter 2

# Production of Bioactive Phytochemicals from Ethnomedicinal Plant Endophytic Microorganisms

*Jacqueline K. Makatiani and Elijah Kipkorir Lelmen*

## Abstract

Ethnomedicinal plants are extensively explored for the discovery of new bioactive phytochemicals with applications in agrochemical, food and pharmaceutical industries. This could lead to extinction of some endangered plant species, hence the need for new phytochemical production strategies. Medicinal plant-associated microbial endophytes act as a promising eco-friendly source of high-valued bioactive phytochemicals. Numerous reports have focused on bioprospecting of endophytes for novel or existing biomolecules. However, investigations on novel production techniques of endophytes for high-value-low volume secondary metabolites are extremely limited. Moreover, the exploration of bacterial and actinomycete endophytes as phytochemical sources has not been fully elucidated. This chapter highlights the diversity, distribution, and bioactivity of phytochemicals in ethnomedicinal plant-associated endophytes isolated between 2014 and 2024. Additionally, recent advances in biotechnological exploration and strain improvement techniques such as genome editing of endophytes for enhanced *in vitro* production of bioactive secondary metabolites are discussed.

**Keywords:** endophytes, medicinal plants, biomolecules, bioactive compounds, natural products, gene editing, *in vitro* production

## 1. Introduction

Over the years, ethnomedicinal plants remained an essential source and reservoir of natural bioactive phytochemicals, particularly secondary metabolites used in the production of agrochemicals, food, and also pharmaceuticals for therapeutic purposes [1]. However, their use as raw materials for novel drug sources has been greatly limited because many ethnomedicinal plants are becoming endangered species due to unabated demand and unsustainable harvesting that contributes to decimation of their wild populations. Consequently, this has led to an increase in the use of synthetic materials for drug production. Currently, available synthetic compounds used in drug development and human disease management have fewer therapeutic benefits and cause harmful effects on patients and the environment.

While the need for new therapeutic strategies emphasizing natural and safe bioproducts and technologies is vital, a considerable quantity of medicinal plants is required for the production of high-yield secondary metabolites. Some ethnomedicinal plants that produce most of the high-valued and often demanded bioactive secondary metabolites have a long generation time to reach maturity, while some are strictly seasonal and thus unavailable during the off-season [2, 3]. Besides, the consistent supply of ethnomedicinal plants all year long is limited by the fact that the quality and quantity of naturally synthesized bioactive phytochemicals fluctuate with varying environmental growth conditions that are linked to rapid changes in climate and other factors [4, 5]. In order to satisfy the growing demands for nature-based therapy, there is a need for alternative drug sources that are natural, sustainable, renewable, and also with minimum or no toxicity [6].

Specifically, different approaches have been explored to improve the quality and quantity of bioactive secondary metabolites associated with medicinal plants [6]. One such approach involves bioprospecting for endophytic microorganisms that co-exist with medicinal plants. Endophytic microorganisms are known for their unique ability to influence the quality of bioactive secondary metabolites produced by their hosts. Besides, they are capable of producing high-value bioactive phytochemicals, which are classified as lignans, hormones, alkaloids, isocoumarins, terpenoids, phenylpropanoids, quinones, lactones, phenol and phenolic acids, and aliphatic metabolites among others, that exhibit pharmacological activities similar to those of the host plants [7]. For such reasons, endophytes serve as excellent sources of discovering secondary metabolites for potential development of new drugs and use in other industries [4, 6].

Although most of the research focuses on bioprospecting fungal and bacterial endophytes for novel or existing high-value bioactive phytochemicals such as camptothecin, paclitaxel, podophyllotoxin, and huperzine A [8], very few reports have exploited the novel *in vitro* production techniques of endophytes for large-scale production of high-value secondary bioactive phytochemicals [4, 9]. Besides, the traditional techniques that are used in the production of endophytic metabolites have several constraints that include low metabolite yields from wild-type strains both in the *in vivo* field environments and *in vitro* tissue culture set-ups, the difficulty of cultivating some endophytes in the laboratory due to the widely reported genomic instability of endophytes in axenic culture, and a lack of understanding of many secondary metabolite biosynthetic pathways [4, 10]. Hence, there is a need to enhance the yield and quality of bioactive secondary phytochemicals from endophytes to satisfy the market demand and conserve the already endangered ethnomedicinal plants. Different approaches have been successfully applied in the attempt to enhance secondary metabolite production by endophytes, and these include molecular methods, elicitation, optimization, co-culture, and epigenetic modification. Improving the yield and quality of bioactive compounds from endophytic fungi is crucial for meeting market demand. Therefore, this chapter highlights the diversity, distribution, and bioactivity of endophytes associated with ethnomedicinal plants. It also highlights on recent advances in biotechnological exploration and endophyte strain improvement techniques such as genome editing, gene knockout, heterologous expression, and the classical mutagenesis, together with techniques for unlocking the endophytic silent gene clusters such as co-cultivation and epigenetic modifications for enhanced *in vitro* production of bioactive secondary metabolites that have been used in the last decade, 2014 and 2024.

## 2. Diversity and distribution of ethnomedicinal plant endophytes

Endophytic microorganisms (fungi, bacteria, and actinomycetes) inhabit the intercellular spaces of healthy plant root tissues (below ground) or phyllosphere (above-ground) tissues, including phylloplane (leaves), anthosphere (fowers), carposphere (fruits), and spermosphere (seeds) [11], without causing any known detrimental effect or symptoms of disease in the host plant, but are useful to the host for a number of reasons [12]. Endophytes occurring in ethnomedicinal plants are reported as the most useful and greatly diverse group of microbiota isolated from plants of the different ecosystems such as rainforests, tropical forests, coastal forests, mangrove swamps, Arctic, Antarctic, geothermal lands, and deserts [13].

The community of endophytic microorganisms that can colonize a medicinal plant throughout its life cycle is influenced by environmental factors, which shape their species composition, diversity, distribution, functions, and community structure [14]. In comparison to medicinal plants from temperate regions, the endophytes in the arid regions have been shown to exhibit less diversity [15]. A review in [13] showed that endophyte diversity in medicinal plants is not only influenced by environmental conditions, but also by host plant characteristics that include genotype, type of tissue or organ, developmental stage (age), health status, nature of phytochemicals present in the plant, and geographic distribution.

Araújo et al. [16] reported high fungal species diversity and abundance in the stems and roots, and leaves, respectively, of the Rubberwood *Hevea brasiliensis* growing in the Brazil Amazon forest. However, in Ref. [17], the *Gentiana officinalis* plant showed that fungal species diversity and abundance were highest in stem than in the leaves and roots, while the bacterial species diversity and richness were highest in the roots than in stem and leaves thus indicating that fungal and bacterial endophytes show a degree of plant species and organ specificity. In [18], it was also reported that it was plant organs rather than the habitat that determined the microbial communities of buds and twigs in young beech medicinal *Fagus sylvatica* forests. Additionally, tree organs of *Alnus incana* and *Corylus avellana* in hemiboreal forest greatly influenced the fungal community structure, with the trunk showing distinct variations from the leaves and branches [19]. Nonetheless, host species and locality had no significant impact, thus indicating that the fungal community is less affected by the external environment [17].

Additionally, the abundance level and frequency of *Phoma* species from various organs of *Caralluma acutangula*, *Rhazya stricta*, and *Moringa peregrina* medicinal plants collected from the arid regions of the Sultanate of Oman were significantly higher than those of the genera *Alternaria*, *Bipolaris*, or *Cladosporium* [15]. Su et al. [20] also found that colonization rates of endophytic fungi of the *Stipa grandis* in the Inner Mongolia steppe were significantly higher in the roots as compared to the leaves and that both tissues also showed distinct variations in the composition and dominant endophytic populations.

Likewise, the structures of endophytic Proteobacteria and Actinobacteria in the leaf and stem tissues of *Mirabilis himalaica* collected from different geographic locations were different compared to those in the root tissues [21]. This mainly results from the differences in the unique internal microenvironment conditions (such as oxygen content, enzymes, chemical components, and nutrients) of different tissues and organs, resulting in a specific selection of survival sites by endophytic communities [22]. Additionally, comparisons of the endophytic fungi present in various tissues and organs of *Polygala crotalarioides* plants growing in Yunnan Province, China, showed a higher fungi abundance in the two-year-old plant than the 1-year-old plants [23]. Thus indicating that the distribution and population structure of endophytic fungal communities depends on the plant age [17].

Endophytes have been isolated from a variety of ethnomedicinal plant species including woody trees, shrubs, and herbs. Generally, the most common genera isolated belong to Ascomycota (e.g., the genera *Alternaria*, *Colletotrichum*, *Epicoccum*, *Fusarium*, *Gibberella*, *Glomerella*, *Guignardia*, *Leptosphaerulina*, *Nigrospora*, *Phoma*, *Phomopsis*, *Tubercularia*, and *Xylaria*), while a few belong to Basidiomycota (Genus *Trametes*), and Zygomycota (Genus *Umbelopsis*) [24]. Majority of the bacterial (Proteobacteria, Bacteroidetes, and Actinobacteria) endophytes isolated belong to the genera *Bacillus*, *Pantoea*, *Pseudomonas*, *Stenotrophomonas*, and *Serratia*. To date, various bioactive secondary metabolites that are produced by endophytes include: Camptothecin, Cladosporin, Cytosporin D, Deacetylmycoepoxydiene, Digoxin, Fengycin, Hypericin, Hypocrellin A, Lovastatin, Paclitaxel, Palmarumycins, Pinoresinol monoglucoside, Podophyllotoxin, Quercetin, Tanshinone IIA, Taxol, Valinomycin, Vinblastine, Vincamine, and many others [7, 9]. These bioactive compounds have been broadly used in antibiotic and anticancer drugs; they shield their host plants from adverse environmental stressors and bio-agents and also play a role in siderophore production, phosphate solubilization, and atmospheric nitrogen fixation [25].

## 3. Yield enhancement strategies for secondary metabolites of ethnomedicinal plant endophytes

Substantial efforts to isolate and screen endophytes for high-value metabolites have previously resulted in low yields of the target metabolite. Co-integration of several scientific tools and approaches such as mutagenesis, genetic transformations, optimization of fermentation parameters, elicitor/inhibitor addition, precursor feeding, use of adsorbent resins/solid supports, co-cultivation and mixed fermentation, and use of epigenetic modifications may be used to effectively target and enhance the secondary metabolite yield and productivity of endophytes. This section highlights the approaches that have been successfully applied in the attempt to enhance biosynthesis of secondary metabolites in endophytic microorganisms isolated from ethnomedicinal plants in the last decade (2014–2024). One approach is strain improvement that comprises (i) Classical random or rational mutagenesis that is occasionally combined with protoplast fusion techniques and (ii) Genetic engineering that involves gene cluster stitching for reconstruction of secondary metabolite biosynthetic metabolic pathways and genome editing of programmable nucleases [6]. The other approach entails techniques for unlocking the endophytic silent gene clusters such as co-cultivation and epigenetic modifications.

### 3.1 Strain improvement

The process of strain improvement in endophytes plays a key role in the *in vitro* production of many secondary metabolites because it genetically modifies microbial strains. The aim is not only for product yield enhancement but also improvement of other characteristics including removal of unwanted co-metabolites, improved nutrient utilization, and modifying cellular morphology for improved oxygen transfer during fermentation. Besides, it also reduces the fermentation time, increases tolerance to high product concentrations, and supports high production of native or foreign products through genetic recombination [6]. Classical mutagenesis and genetic engineering are the two approaches used for strain improvement (**Table 1**).

### *3.1.1 Mutagenesis of endophytic microorganisms*

Classical strain improvement technique has long been regarded as the gold standard for strain improvement and enhancing the production of microbial metabolites that are of commercial significance. Application of this technique does not require an in-depth understanding of the genetic basis of the manipulated microorganisms or their biosynthetic pathways [6]. The use of one mutagen repeatedly or two or more mutagens, either chemical (nitrosoguanidine [NTG], diethyl sulfate, ethyl methyl sulfonate [EMS], 1-methyl-3-nitro-1-nitrosoguanidine [MNNG], sodium nitrite [NaNO2], nitrous acid [HNO2] and diethyl sulfate [DES]) or physical (ultraviolet radiation [UV], X-rays, gamma radiation [γ-rays], laser, ion beams and microwave) changes in the genetic makeup of microbial strains. These changes result in mutations in the microbial genome that can also increase the synthesis of secondary metabolites [32]. Generally, the selection of improved mutants is often done through random screening for the desired trait and relies on exposure to a specific mutagen with the aim of inserting point mutations at random in the organism's genome (**Table 1**) [26–39, 46]. Alternatively, rational screening involves the selection of specific mutants with desired genotype characteristics that vary from the previous genotype of interest [28].

In endophytic fungi, mutagenesis is done using mycelia, spores, or protoplasts. However, mutation of endophyte mycelium highlights the demerits of genetic divergence in the offspring, while optimization of conditions for endophyte spore mutagenesis is often difficult. Previous successful studies [47, 48] on mutagenesis of endophytes for strain improvement have focused on the optimization of fermentation parameters and also the protocols for protoplast preparations. This suggests that protoplast fusion is the best way to induce mutagenesis [49]. Protoplast fusion allows the genetic recombination between genomes of different parental strains, followed by selection of hybrid strains showing the desired traits [50]. It requires specific enzymatic hyrolyzation of protoplasts to remove the cell wall, followed by genetic exchange and recombination via cytoplasmic and nuclear fusion and the regeneration of the cell wall that subsequently gives rise to hybrid cells of the desired shuffled strain [51].

A few reports have discussed the yield enhancement of the targeted metabolite via protoplast mutagenesis and regeneration (**Table 1**) [30, 34, 49, 52–54], highlighting its beneficial outcomes. Successful strain improvement by induction of genetic variabilities via mutations in endophytic fungi *Aspergillus luchuensis* derived from marine mangrove showed that lovastatin yield increased 32-fold (87.0 mg/gds) by the recombinant strain [26]. In this case, fungal conidia was first treated with a combination of ultraviolet radiation (UV) followed by N-methyl-N-nitro-N-nitrosoguanidine (NTG) mutagens, then with a second combination of NTG and ethidium bromide, followed by protoplast fusion with three successive rounds of gene shuffling. Lovastatin is a natural inhibitor of

| Approach and method | Endophyte | Medicinal plant | Secondary metabolite | Improved yield | Ref. |
|---|---|---|---|---|---|
| Mutagenesis<br>• UV + NTG<br>• NTG + EtBr<br>• Gene shuffling | *Aspergillus luchuensis* | Marine mangrove | Lovastatin | 87.0 mg/gds (32-fold) | [26] |
| Mutagenesis<br>• UV + NTG<br>• Gene shuffling | *Phomopsis species* | Mangrove plants | Deacetylmyco-epoxydiene (DAM) | 219 mg/L (275-fold) | [27] |
| Mutagenesis<br>• Protoplast inactivation<br>• UV + heat<br>• Gene shuffling | *Emericella foeniculicola* | *Salvia miltiorrhiza* | Tanshinone IIA | 387.56 mg/g (11.07-fold) | [28] |
| Mutagenesis<br>• γ-irradiation | *Shiraia* species | *Phyllostachys edulis* | Hypocrellin A | 2018.3 mg/L (414.9% increase) | [29] |
| Mutagenesis<br>• Inactivated protoplast fusion + UV<br>• + $NaNO_2$ | *Tausonia pullulans* | *Vinca minor* | Vincamine | 31.6 mg (162.24% increase)<br>38.7 mg (221.16% increase) | [30] |
| Mutagenesis<br>• γ-irradiation | *Alternaria alternata* | *Melissa officinalis,* | Vinblastine | 817.44 µg/L (3.98-fold) | [31] |
| Mutagenesis<br>• UV + $HNO_2$ + EtBr | *Aspergillus* species | — | Taxol | 99.44 mg/L (14% increase) | [32] |
| Mutagenesis<br>• γ-irradiation | *Epicoccum nigrum* | *Terminalia arjuna* | Digoxin | 10.65 mg/L (5-fold) | [33] |
| Mutagenesis<br>• UV + protoplast + elicitor (*T. baccata* bark) | *Acremonium* species | *Taxus baccata* | Paclitaxel | 160.85 µg/L (2.5-fold) | [34] |

| Approach and method | Endophyte | Medicinal plant | Secondary metabolite | Improved yield | Ref. |
|---|---|---|---|---|---|
| Mutagenesis<br>• γ-irradiation | *Epicoccum nigrum* | *Taxus baccata* | Taxol | 40.07% increase (1.66-fold) | [35] |
| Mutagenesis<br>• UV | *Aspergillus fumigatus*<br>*Alternaria tenuissima* | *Taxodium distichum*<br>*Terminalia arjuna* | Paclitaxel | 504.56 µg/L (1.16-fold)<br>195.64 µg/L (1.20-fold) | [36] |
| Mutagenesis<br>• γ-irradiation | *Aspergillus fumigatus*<br>*Alternaria tenuissima* | *Taxodium distichum*<br>*Terminalia arjuna* | Paclitaxel | 495.31 µg/L<br>195.41 µg/L | [36] |
| Mutagenesis<br>• γ-irradiation<br>• γ-irradiation+ nutrition optimization | *Penicillium polonicum* | *Ginkgo biloba* | Taxol | 401.2 mg/L (1.2-fold)<br>401.2 mg/L (4.4-fold) | [37] |
| Mutagenesis<br>• UV + RE + fermentation optimization | *Streptomyces flavoviridis*<br>*Endophytic actinomycetes* | — | 6′-deoxy-bleomycin Z | 70.7 mg/L (7-fold) | [38] |
| Mutagenesis<br>• NTG + antibiotics<br>• Gene shuffling | *Bacillus amyloliquefaciens* | *Scutellaria baicalensis* | Fengycin | 450.51 mg/L (8.30- fold) | [39] |
| Genetic transformation<br>• CRISPR/Cas9 system Gene editing<br>• 1, 3, 6, 8-tetrahydroxy-naphthalene synthase gene + optimization | *Berkleasmium* species | *Dioscorea zingiberensis* | Palmarumycins | — | [40] |

| Approach and method | Endophyte | Medicinal plant | Secondary metabolite | Improved yield | Ref. |
|---|---|---|---|---|---|
| Genetic transformation<br>• Modified CRISPR/Cas9 mediated by double gRNA sequences<br>• MAPKK gene knock out | *Phomopsis liquidambaris* | *Bischofia polycarpa* | Flavonoids;<br>Naringenin<br>Kaempferol<br>Quercetin | 5.65 mg/L (2.3-fold)<br>1.96 mg/L (3.4-fold)<br>2.37 mg/L (3.4-fold) | [41] |
| Genetic transformation<br>• Hygromycin-resistance Gene- Knockout + Gene Retro-Complementation with plasmid vector and selected marker | *Colletotrichum gloeosporioides* | *Huperzia serrata* | Huperzine A | Significant yield increase | [42] |
| Genetic transformation<br>• Gene deletion | *Streptomyces species* | *Citrus sinensis* | Valinomycin | 5.171 mg/L (10.5-fold) | [43] |
| Genetic transformation<br>• Genetic engineering<br>• Amplification of Pscrt biosynthetic gene cluster + heterologous expression | *Pseudomonas* species | *Taxus chinensis* | Zeaxanthin diglucoside | 380 mg/L (85% increase) | [44] |
| Genetic transformation<br>• Genetic engineering<br>• taxadiene-biosynthetic cluster + Heterologous expression<br>• In *Escherichia coli* | *Alternaria alternata* | *Taxus chinensis* var. *Mairei* | Taxadiene | 61.9 μg/L | [45] |

*UV—Ultraviolet radiation; NTG—N-methyl-N′-nitrosoguanidine; EtBr—Ethidium bromide; γ—Gamma irradiation; $HNO_2$—Nitrous acid; $NaNO_2$—Sodium nitrite; RE—Ribosome engineering; Pscrt—Pseudomonas species carotenoid biosynthetic gene cluster; MAPKK—Mitogen-activated protein kinase kinase; CRISPR—clustered regularly interspaced short palindromic repeats; Cas9—CRISPR-associated protein 9.*

**Table 1.**
*Summary of studies on strain improvement of endophytes associated with etnomedicinal plants.*

3-hydroxy-3-methylglutarylcoenzyme A reductase enzyme, and it is used as an anti-cholesterol drug molecule [26]. Also, a mutant strain obtained via genome shuffling of the parental protoplasts of the endophytic fungus *Phomopsis* species yielded 275-fold (219 mg/L) increase of the antitumor metabolite deacetylmycoepoxydiene [27]. Furthermore, attempts of subjecting high tanshinone IIA-producing mutant fungal strains of *Emericella foeniculicola* (Aspergillaceae: *Aspergillus*) to genome shuffling and three rounds of protoplast fusion resulted in an improved recombinant strain that produced 387.56 ± 0.02 mg/g of tanshinone IIA which was 11.07 times higher when compared to the parental strain [28].

The fungal wild strain was isolated from *Salvia miltiorrhiza*, a well-known traditional Chinese herbal medicine. It is one of the best natural pigments and bioactive agents for treating skin diseases. When the spores of a Hypocrellin A (HA)-producing strain (*Shiraia* species) that was isolated from *Phyllostachys edulis* seeds were mutated by gamma irradiation, HA production in the mutant strain reached 2018.3 mg/L (414.9% increase) when compared to that of the parent strain (392 mg/L) [29], a level that was significantly higher than all the other industrial HA-producing strains in published reports. Similarly, improvement in the production of vincamine in endophytic fungus *Vinca minor* (VINI-7), isolated from *Tausonia pullulans*, by using the inactivated protoplast fusion method was reported in [30]. Vincamine, an indole alkaloid, is widely found in plants of the Apocynaceae family and has significant effects on cell proliferation and on cardiovascular, cerebrovascular, and nervous system diseases. Vincamine yield in fusants inactivated by UV radiation was enhanced to 31.6 mg (162.24% increase), while yield in fusants inactivated by sodium nitrite was enhanced to 38.7 mg (221.16% increase). Under optimized fermentation conditions, a mutant strain of an endophytic actinomycete *Streptomyces flavoviridis* yielded a sevenfold (70.7 ± 1.8 mg/L) increase in 6′-deoxy-bleomycins compound after the parental strain was subjected to a combination of UV mutagenesis and ribosome engineering [38]. These results indicate the development of classical mutation and selection strategies as a promising approach to improving secondary metabolite production by endophytes. Other examples of successful use of mutagenesis to enhance secondary metabolite yield in endophytes derived from ethnomedicinal plants are shown in **Table 1**.

#### *3.1.2 Genetic transformations in endophytes*

Unlike mutagenesis which does not require specific genetic knowledge of the manipulated microorganisms, genetic transformation requires precise control of the manipulations and modifications introduced in an organism's genome [6]. It comprises of alteration in the genome sequence or exogenous DNA insertion into a cell to activate cryptic genes, which allows increasing metabolite production levels, producing novel compounds or directing the synthesis of the desired products [40–45, 55]. Genetic manipulation of specific endophytic microorganisms still presents a challenge, since there is still lack of developed genetic transformation tools for many microbial species [56]. A review in [57] indicates that most of the studies on the genetic transformation of endophytes are still at a preliminary stage, focusing on the development of efficient transformation methods and expression tools. Nevertheless, genetic transformations in representatives of all major fungal species are now possible. Thus, DNA transformations aimed at establishing efficient biosynthesis of gene products have great potential for use in agrochemical, food, and pharmaceutical industries [58].

The commonly used methods for genetic transformation in fungi are protoplast-mediated transformation, *Agrobacterium tumefaciens*-mediated transformation (ATMT), and electroporation [59]. Heterologous expression is one of the useful approaches that allow one species to express foreign proteins that they naturally do not have but are derived from the introduction of a foreign gene from another species, encoding for a protein of interest [60]. Generally, the specific target gene is transformed using a vector, then what follows is culturing and expressing, identifying the chemical structure, and confirming its bioactivities. Host organisms that are mainly used for heterologous expression are *Escherichia coli* (bacteria), *Saccharomyces cerevisiae* (yeast), and the filamentous fungi *Aspergillus nidulans* and *Aspergillus oryzae* for obtaining fungal secondary metabolites [61]. Recently, a procedure for Polyethylene Glycol (PEG)-mediated transformation of the taxol-producing endophytic fungi *Ozonium* species isolated from *Taxus chinensis* var. *mairei* was established [62], where protoplasts obtained by the optimized enzyme system and Plasmid pV2 containing the hygromycin-B phosphotransferase gene driven by a fungal promoter (trpC) was successfully integrated into the fungal genome. The establishment of such an efficient transformation system of taxol-producing endophytic fungus offers great potential for improving taxol production by engineering the taxol biosynthetic pathway genes in the future. Examples of successful use of genetic transformation to enhance secondary metabolite yield in endophytes derived from ethnomedicinal plants are shown in **Table 1**.

In a report of pathway reconstitution, the taxadiene-biosynthetic cluster was expressed in *Escherichia coli* and transformed into endophytic fungus *Alternaria alternata* isolated from *T. chinensis* var. *mairei*, and the mevalonate (MVA) pathway was engineered, with introduction of the plant taxadiene-forming gene [45]. This resulted in the stable production of 61.9 ± 6.3 μg/L of taxadiene in A. alternata (**Table 1**), thus providing the first demonstration of taxadiene production in filamentous fungi and an alternative method for Taxol production in microorganisms. Taxadiene is the first committed precursor in the biosynthesis of Taxol.

In another report, a carotenoid (*Pscrt*) biosynthetic gene cluster from the endophytic bacterium *Pseudomonas* species that was isolated from the leaves of *T. chinensis* was amplified, and then successfully reconstituted in heterologous hosts *E. coli* and *Pseudomonas putida* to enhance the production of a promising antioxidant zeaxanthin diglucoside [44] in *Pseudomonas* species. The production was improved through optimization of fermentation conditions (medium, cultivation temperature, and culture time) and a further enhancement by the introduction of an expression plasmid harboring the Pscrt biosynthetic gene cluster into *Pseudomonas* species. This resulted into the engineered strain producing 380 ± 12 mg/L of zeaxanthin diglucoside, a level that was 85% higher than production in the wild type. This engineered strain represents a great host for the industrial production of this promising glycosylated carotenoid.

Another cutting-edge technology, genome editing, has great potential to revolutionize the production of natural products. The clustered, regularly interspaced short palindromic repeats (CRISPR)/Cas9 system is a programmable genome editing technology that allows for precise and efficient modifications to the DNA of all cell types [63]. The system forms an active Cas9/guide RNA complex that requires a guide RNA to match a desirable microbial gene of interest that ensures efficient protein translation and a Cas9 (CRISPR-associated protein 9) endonuclease that makes double-stranded breaks in genomes, thus allowing particular genes associated with the biosynthesis of pharmacological important chemicals to be deleted, added, or modified [63, 64].

Recently, numerous protocols for the use of CRISPR/Cas9 system in several species of fungi have been developed [65]. As reviewed in [57], this technology has been applied to a limited number of real applications in model filamentous fungi such as *Sporormiella minima*, *Aspergillus oryzae*, *Aspergillus niger*, *Aspergillus nidulans*, *Talaromyces atroroseus*, *Neurospora crassa*, *Alternaria alternata*, and *Penicillium chrysogenum* to explore increased production of biologically active compounds. Therefore, development of these systems in endophytic fungi has not been fully elucidated. Most reports focus on these system's feasibility in different microbial species, and additionally, there are limited review articles available on this topic [66, 67].

A study in Ref. [41] showed that CRISPR-Cas9-based genome editing in the endophytic fungus *Phomopsis liquidambaris* isolated from the inner bark of the medicinal plant *Bischofia polycarpa* produces the flavonoids naringenin, kaempferol, and quercetin. *Phomopsis liquidambaris* mitogen-activated protein kinase kinase (MAPKK) signaling gene not only is important for mutualistic symbiosis with plants [68], but also affects the synthesis of secondary metabolites. The MAPKK gene was successfully knocked out via double gRNA introduction into a newly constructed CRISPR/Cas9 system, which was conducive to causing plasmid loss [41]. This triggered the biosynthesis of Naringenin, kaempferol, and quercetin yielded 5.65 mg/L, 1.96 mg/L, and 2.37 mg/L, respectively.

Nonetheless, there is a recent study on the development of a CRISPR/Cas9 system suitable for gene editing in the endophytic fungus *Berkleasmium* species that were isolated from *Dioscorea zingiberensis*, and produces palmarumycins (polyketides of the spirobisnaphthalene class) those exhibit antibacterial, antifungal, and cytotoxic activities [40]. Fungal protoplast preparation was optimized before disrupting the *bdpks* gene that encodes the pentaketide 1,3,6,8-tetrahydroxynaphthalene synthase enzyme, which is responsible for the biosynthesis of 1,8-dihydroxynaphthalene-melanin. The successful gene disruption provided a platform for the construction of an optimal system using the U6 snRNA-1 promoter as the sgRNA promoter that resulted in the termination of spirobisnaphthalenes and melanin production. Hence, this gene disruption platform forms a basis for further exploration of other gene functions and also for the development of efficient CRISPR/Cas9 systems in other endophytes [40].

Gene deletion or gene knockout strategies have also enabled the generation of strains with improved production of high-value secondary metabolites and inhibition of toxic metabolites in filamentous fungi by metabolic engineering. Generally, one target gene is made inoperative in an organism. Deleting a gene completely usually requires double crossover homologous recombination events at the genomic level [69]. The genus *Streptomyces* represents one of the largest producers of molecules with antibiotic activity. From a study conducted in [43], the whole genome sequencing of the endophytic microorganism *Streptomyces* species that produces valinomycin was isolated from *Citrus sinensis* branches and revealed 35 gene clusters encoding for secondary metabolism. Valinomycin plays several biological functions such as antibacterial, antiviral, and anticancer [69]. In the study, deleting an adenylation domain from a hybrid non-ribosomal peptide synthetases (NRPS) cluster through a double crossing knockout experiment resulted in the mutant strain producing 5.171 ± 1.048 mg/L of valinomycin, an increased production of 10.5-fold (**Table 1**).

### 3.2 Unlocking the cryptic gene clusters in endophytes

Silent biosynthetic genes or genomic instability results from the application of bioprocess optimization strategies on endophytic fungi for enhanced product yields in axenic culture. As depicted in **Table 2**, the co-cultivation of compatible endophytic strains and the use of epigenetic modifiers to induce the production of secondary metabolites by endophytes are approaches that have great potential to overcome the shortfalls associated with stabilizing the productive capabilities of axenic cultures [4, 25, 70–72].

### *3.2.1 Co-cultivation of endophytes*

Several novel compounds have been isolated through co-cultivation method. Co-cultivation is the combination of two or more microbial strains together to enhance the product yield of known secondary metabolites and also stimulate the biosynthesis of novel secondary metabolites that are often absent in traditional single cultures [73]. Although the exact mechanism of secondary metabolite biosynthesis remains unclear in many endophytic co-cultures, there is indication of signaling molecules that one strain produces are exogenic to the other strain and hence can awaken the silent or cryptic biosynthetic gene clusters to produce new compounds [74]. In this method, an artificial microbial community is constructed to activate the biosynthetic gene clusters to produce new compounds. The co-culture method involves several strategies that include co-culturing: (a) two endophytic fungi; (b) endophytic fungi with bacterium; or (c) endophytic fungi with fungi isolated from diverse habitats that make enhancement or production of new compound and activation of silent gene clusters for microorganisms [75].

In a previous study [70], co-culture of endophytic fungus *Phomopsis* species isolated from the bark of medicinal herb *Eucommia ulmoides* and fungus *Alternaria* species isolated from grape, and co-culture of *Phomopsis* sp. and *E. ulmoides* bark increased the yield of two natural lignans Pinoresinol and (+)-pinoresinol monoglucoside (PMG) by 2.57- and 1.18-fold, respectively, (**Table 2**) in liquid potato dextrose broth medium (PDB) and 1.25- and 1.52-fold, respectively, in cell medium with only glucose. Pinoresinol and PMG are enterolactone precursors that can prevent breast

| Approach & method | Endophyte | Medicinal plant | Secondary metabolite | Improved yield | Ref. |
|---|---|---|---|---|---|
| Co-cultivation<br>• *Alternaria sp. + Eucommia ulmoides bark in the medium* | *Phomopsis* species | *Eucommia ulmoides* | Pinoresinol<br>Pinoresinol monoglucoside<br>Pinoresinol<br>Pinoresinol monoglucoside | 2.57-fold in PDB<br>1.18-fold in PDB<br>1.25-fold resting cells<br>1.52-fold in resting cells | [70] |
| Epigenetics<br>• valproic acid | *Aspergillus fumigatus* | *Grewia asiatica* | Fumiquinazoline C | 49.30 μg/mg (10-fold) | [71] |
| Epigenetics<br>• 5-azacytidine | *Penicillium radicina* | *Salvia abrotanoides* | Tanshinone IIA | 63.176 mg/g | [72] |

*(PDB—Potato Dextrose Broth medium).*

**Table 2.**
*Summary of studies on unlocking gene clusters in endophytes associated with etnomedicinal plants.*

and endometrial cancers. Elicitation of lignan synthesis and the expression of the silent lignan gene clusters can be attributed to the microbial interactions in the co-cultures that consequently resulted in enhanced Pinoresinol and PMG yields [70].

### *3.2.2 Application of epigenetic modifiers*

Epigenetics focus on alterations in the cell or organism's genomic structure and pattern of gene expression without changing the DNA sequences. Thus, those alterations can influence the kind of proteins that are transcribed by switching genes on or off [76]. Generally, the full biosynthetic potential of endophytic microorganisms is extremely limited because most of their gene clusters that encode bioactive secondary metabolites are often rendered silent and cryptic under axenic cultures. However, their gene expression can be altered without any modification in the order of nucleotides in DNA. Epigenetic modifications using chemical inhibitors (more specifically, Histone deacetylase [HDAC] inhibitors such as sodium butyrate, suberoylanilide hydroxamic acid, and valproic acid or DNA methyltransferase [DNMT] inhibitors such as 5-azacytidine, hydralazine, and procaine) or inducers are capable of awakening the expression of silent biosynthetic pathways in endophytic microorganisms, hence resulting in enhanced yields of bioactive secondary metabolites [71, 72, 77].

Fumiquinazolines are peptidyl alkaloids and are reported to have significant antibacterial, antifungal, and antitumor properties [78]. A study on the effect of valproic acid on the biosynthetic pathway of fumiquinazoline C in endophytic fungi *Aspergillus fumigatus* isolated from *Grewia asiatica* showed ten folds enhancement in the production of a secondary metabolite, fumiquinazoline C (49.30 μg/mg crude extract) from the initial quantities (4.61 μg/mg crude extract) [71]. Nonetheless, those findings also showed that endophytic exposure to valproic acid triggered up-regulation of all the genes, that is, Afua_6g 12,040, Afua_6g 12,050, Afua_6g 12,060, Afua_6g 12,070 and Afua_6g 12,080 responsible for the biosynthesis of fumiquinazoline C. This indicates that epigenetic modifiers can tickle the biosynthetic pathway, resulting in an alteration in the metabolic profile of the microorganisms [71].

The 5-azacytidine (5-AC) is an analog of cytidine nucleoside molecule, and its incorporation into DNA often inhibits DNA methylation; hence, it can induce the expression of silenced secondary metabolite genes in fungi [72]. In a recent study highlighting the first case of applying 5- azacytidine in endophytic fungi for enhanced tanshinone yield, this epigenetic modifier restored the production of tanshinone IIA by a threefold increase in the third subculture of endophytic fungi *Penicillium pinophilum* isolated from the roots of *Salvia abrotanoides* [72]. Similarly, the same 5-AC epigenetic modifier increased Tanshinone IIA production in the endophytic fungal cultures of *Penicillium radicina* (63.176 mg TIIA/g of dry weight) and *Penicillium nodositatum*. These findings confirmed 5- azacytidine as an efficient, easy, and quick chemical to felicitate secondary metabolite production in endophytic fungi [72].

## 4. Conclusion

*In vitro* strategies show great promise in improving the yields of target bioactive phytochemicals from endophytes, which can be explored to elicit the synthesis of desired natural molecules for use in agrochemical, food, and pharmaceutical industries. Therefore, promising ethnomedicinal plant endophytic strains can be manipulated for enhanced production of high-value phytochemical compounds through mutagenesis that causes genetic changes that can increase the synthesis of secondary metabolites, genome editing tools that modulate gene expression and co-cultivation, and epigenetic modification that enhance production of novel compounds, and activates the silent and cryptic gene clusters in endophytes that code for the production of secondary metabolites.

## Acknowledgements

The authors wish to thank the Department of Biological Sciences, Moi University, and the Department of Biochemistry and Molecular Biology, Egerton University, Kenya.

## Conflict of interest

The authors declare no conflict of interest.

## Author details

Jacqueline K. Makatiani[1,2]* and Elijah Kipkorir Lelmen[3]

1 Department of Biological Sciences, Moi University, Eldoret, Kenya

2 Africa Center of Excellence in Phytochemicals, Textile and Renewable Energy, Moi University, Eldoret, Kenya

3 Department of Biochemistry and Molecular Biology, Egerton University, Egerton, Nakuru, Kenya

*Address all correspondence to: jkmakatiani@mu.ac.ke

**IntechOpen**

© 2024 The Author(s). Licensee IntechOpen. This chapter is distributed under the terms of the Creative Commons Attribution License (http://creativecommons.org/licenses/by/4.0), which permits unrestricted use, distribution, and reproduction in any medium, provided the original work is properly cited.

## References

[1] Kouipou Toghueo RM, Boyom FF. Endophytic fungi from *Terminalia* species: A comprehensive review. Journal of Fungi. 2019;**5**(2):43. DOI: 10.3390/jof5020043

[2] Verma N, Shukla S. Impact of various factors responsible for fluctuation in plant secondary metabolites. Journal of Applied Research on Medicinal and Aromatic Plants. 2015;**2**(4):105-113. DOI: 10.1016/j.jarmap.2015.09.002

[3] Isah T. Stress and defense responses in plant secondary metabolites production. Biological Research. 2019;**52**:39. DOI: 10.1186/s40659-019-0246-3

[4] Venugopalan A, Srivastava S. Endophytes as *in vitro* production platforms of high value plant secondary metabolites. Biotechnology Advances. 2015;**33**(6):873-887. DOI: 10.1016/j.biotechadv.2015.07.004

[5] Shasmita SNR, Rath SK, Behera S, Naik SK. *In vitro* secondary metabolite production through fungal elicitation: An approach for sustainability. In: Prasad R, Kumar V, Kumar M, Wang S, editors. Fungal Nanobionics: Principles and Applications. Singapore: Springer; 2018. pp. 215-242. DOI: 10.1007/978-981-10-8666-3_9

[6] Tsipinana S, Husseiny S, Alayande KA, Raslan M, Amoo S, Adeleke R. Contribution of endophytes towards improving plant bioactive metabolites: A rescue option against red-taping of medicinal plants. Frontiers in Plant Science. 2023;**14**:1248319. DOI: 10.3389/fpls.2023.1248319

[7] Ezeobiora CE, Igbokwe NH, Amin DH, Mendie UE. Endophytic microbes from Nigerian ethnomedicinal plants: A potential source for bioactive secondary metabolites—A review. Bulletin of the National Research Centre. 2021;**45**:103. DOI: 10.1186/s42269-021-00561-7

[8] Tiwari P, Bae H. Endophytic fungi: Key insights, emerging prospects, and challenges in natural product drug discovery. Microorganisms. 2022;**10**(2):360. DOI: 10.3390/microorganisms10020360

[9] Sharma H, Rai AK, Dahiya D, Chettri R, Nigam PS, et al. Exploring endophytes for *in vitro* synthesis of bioactive compounds similar to metabolites produced *in vivo* by host plants. AIMS Microbiology. 2021;7(2):175-199. DOI: 10.3934/microbiol.2021012

[10] Zhang X, Hindra EMA. Unlocking the trove of metabolic treasures: Activating silent biosynthetic gene clusters in bacteria and fungi. Current Opinion in Microbiology. 2019;**51**:9-15. DOI: 10.1016/j.mib.2019.03.003

[11] Adeleke RA, Nunthkumar B, Roopnarain A, Obi L. Applications of plant–microbe interactions in agro-ecosystems. In: Kumar V, Prasad R, Kumar M, Choudhary D, editors. Microbiome in Plant Health and Disease. Singapore: Springer; 2019. pp. 1-34. DOI: 10.1007/978-981-13-8495-0_1

[12] Toppo P, Jangir P, Mehra N, Kapoor R, Mathur P. Bioprospecting of endophytic fungi from medicinal plant *Anisomeles indica* L. for their diverse role in agricultural and industrial sectors. Scientific Reports. 2024;**14**:588. DOI: 10.1038/s41598-023-51057-5

[13] Yadav G, Meena M. Bioprospecting of endophytes in medicinal plants of

Thar Desert: An attractive resource for biopharmaceuticals. Biotechnology Reports. 2021;**30**:e00629. DOI: 10.1016/j.btre.2021.e00629

[14] Walitang DI, Kim CG, Kim K, Kang Y, Kim YK, Sa T. The influence of host genotype and salt stress on the seed endophytic community of salt-sensitive and salt-tolerant rice cultivars. BMC Plant Biology. 2018;**18**:51. DOI: 10.1186/s12870-018-1261-1

[15] Khan AL, Gilani SA, Waqas M, Al-Hosni K, Al-Khiziri S, Kim YH, et al. Endophytes from medicinal plants and their potential for producing indole acetic acid, improving seed germination and mitigating oxidative stress. Journal of Zhejiang University-SCIENCE B. 2017;**18**(2):125-137. DOI: 10.1631/jzus.b1500271

[16] Araújo KS, Brito VN, Veloso TGR, de Leite TS, Alves JL, da Hora Junior BT, et al. Diversity and distribution of endophytic fungi in different tissues of *Hevea brasiliensis* native to the Brazilian Amazon forest. Mycological Progress. 2020;**19**(10):1057-1068. DOI: 10.1007/s11557-020-01613-4

[17] Hou Q, Chen D, Wang YP, Ehmet N, Ma J, Sun K. Analysis of endophyte diversity of *Gentiana officinalis* among different tissue types and ages and their association with four medicinal secondary metabolites. PeerJ. 2022;**10**:e13949. DOI: 10.7717/peerj.13949

[18] Siddique AB, Biella P, Unterseher M, Albrectsen BR. Mycobiomes of young beech trees are distinguished by organ rather than by habitat, and community analyses suggest competitive interactions among twig fungi. Frontiers in Microbiology. 2021;**12**:646302. DOI: 10.3389/fmicb.2021.646302

[19] Küngas K, Bahram M, Põldmaa K. Host tree organ is the primary driver of endophytic fungal community structure in a hemiboreal forest. FEMS Microbiology Ecology. 2019;**96**(2):fiz199. DOI: 10.1093/femsec/fiz199

[20] Su YY, Guo LD, Hyde KD. Response of endophytic fungi of *Stipa grandis* to experimental plant function group removal in Inner Mongolia steppe, China. Fungal Diversity. 2010;**43**(1):93-101. DOI: 10.1007/s13225-010-0040-6

[21] Zhang E, Lu Y, Zhao R, Yin X, Zhang J, Yu B, et al. Endophytic bacterial community structure and diversity of the medicinal plant *Mirabilis himalaica* from different locations. Brazilian Journal of Microbiology. 2023;**54**(4):2991-3003. DOI: 10.1007/s42770-023-01149-1

[22] Wang R, Zhang Q, Ju M, Yan S, Zhang Q, Gu P. The Endophytic fungi diversity, community structure, and ecological function prediction of Sophora alopecuroides in Ningxia, China. Microorganisms. 2022;**10**(11):2099. DOI: 10.3390/microorganisms10112099

[23] Shen K, Xiong Y, Liu Y, Fan X, Zhu R, Hu Z, et al. Community structure and diversity of endophytic fungi in cultivated *Polygala crotalarioides* at two different growth stages based on culture-independent and culture-based methods. Journal of Fungi. 2024;**10**(3):195. DOI: 10.3390/jof10030195

[24] Zheng YK, Qiao XG, Miao CP, Liu K, Chen YW, Xu LH, et al. Diversity, distribution and biotechnological potential of endophytic fungi. Annals of Microbiology. 2015;**66**(2):529-542. DOI: 10.1007/s13213-015-1153-7

[25] Hashem AH, Attia MS, Kandil EK, Fawzi MM, Abdelrahman AS, Khader MS, et al. Bioactive compounds and biomedical applications of endophytic fungi: A recent review.

Microbial Cell Factories. 2023;**22**:107. DOI: 10.1186/s12934-023-02118-x

[26] Ahmed El-Gendy MMA, Al-Zahrani HAA, El-Bondkly AMA. Genome shuffling of mangrove Endophytic *aspergillus luchuensis* MERV10 for improving the cholesterol-lowering agent lovastatin under solid state fermentation. Mycobiology. 2016;**44**(3):171-179. DOI: 10.5941/myco.2016.44.3.171

[27] Wang M, Zhang W, Xu W, Shen Y, Du L. Optimization of genome shuffling for high-yield production of the antitumor deacetylmycoepoxydiene in an endophytic fungus of mangrove plants. Applied Microbiology and Biotechnology. 2016;**100**(17):7491-7498. DOI: 10.1007/s00253-016-7457-0

[28] Zhang P, Lee Y, Wei X, Wu J, Liu Q, Wan S. Enhanced production of tanshinone IIA in endophytic fungi *Emericella foeniculicola* by genome shuffling. Pharmaceutical Biology. 2018;**56**(1):357-362. DOI: 10.1080/13880209.2018.1481108

[29] Liu XY, Shen XY, Fan L, Gao J, Hou CL. High-efficiency biosynthesis of hypocrellin a in *Shiraia* sp. using gamma-ray mutagenesis. Applied Microbiology and Biotechnology. 2016;**100**(11):4875-4883. DOI: 10.1007/s00253-015-7222-9

[30] Xu S, Ren N, Liu J, Wu Y, Yuan G. Improvement of vincamine production of endophytic fungus through inactivated protoplast fusion. International Microbiology. 2020;**23**(3):441-451. DOI: 10.1007/s10123-020-00117-1

[31] El-Sayed ER. Discovery of the anticancer drug vinblastine from the endophytic *Alternaria alternata* and yield improvement by gamma irradiation mutagenesis. Journal of Applied Microbiology. 2021;**131**(6):2886-2898. DOI: 10.1111/jam.15169

[32] Khan R, Mukhtar H, Gohar U, Tahir S. Random mutagenesis of Endophytic fungi for enhanced Taxol production: Enhancement of taxol production by random mutagenesis. Proceedings of the Pakistan Academy Sciences: B Life Environmental Sciences. 2020;**57**(2):67-74

[33] El-Sayed ER, Ahmed AS, Abdelhakim HK. A novel source of the cardiac glycoside digoxin from the endophytic fungus *Epicoccum nigrum*: Isolation, characterization, production enhancement by gamma irradiation mutagenesis and anticancer activity evaluation. Journal of Applied Microbiology. 2019;**128**(3):747-762. DOI: 10.1111/jam.14510

[34] El-Bialy HA, El-Bastawisy HS. Elicitors stimulate paclitaxel production by endophytic fungi isolated from ecologically altered *Taxus baccata*. Journal of Radiation Research and Applied Sciences. 2020 Jan;**13**(1):79-87. DOI: 10.1080/16878507.2019.1702244

[35] El-Sayed ESR, Zaki AG, Ahmed AS, Ismaiel AA. Production of the anticancer drug taxol by the endophytic fungus *Epicoccum nigrum* TXB502: Enhanced production by gamma irradiation mutagenesis and immobilization technique. Applied Microbiology and Biotechnology. 2020;**104**(16):6991-7003. DOI: 10.1007/s00253-020-10712-x

[36] ESR E-S, Ismaiel AA, Ahmed AS, Hassan IA, AZA KE-D. Bioprocess optimization using response surface methodology for production of the anticancer drug paclitaxel by *aspergillus fumigatus* and *Alternaria tenuissima*: Enhanced production by ultraviolet and gamma irradiation. Biocatalysis and Agricultural Biotechnology.

2019;**18**:100996. DOI: 10.1016/j.bcab.2019.01.034

[37] Abdel-Fatah SS, El-Batal AI, El-Sherbiny GM, Khalaf MA, El-Sayed AS. Production, bioprocess optimization and γ-irradiation of *Penicillium polonicum*, as a new Taxol producing endophyte from *Ginko biloba*. Biotechnology Reports. 2021;**30**:e00623. DOI: 10.1016/j.btre.2021.e00623

[38] Zhu X, Kong J, Yang H, Huang R, Huang Y, Yang D, et al. Strain improvement by combined UV mutagenesis and ribosome engineering and subsequent fermentation optimization for enhanced 6′-deoxy-bleomycin Z production. Applied Microbiology and Biotechnology. 2017;**102**(4):1651-1661. DOI: 10.1007/s00253-017-8705-7

[39] Zhao J, Zhang C, Lu J, Lu Z. Enhancement of fengycin production in *bacillus amyloliquefaciens* by genome shuffling and relative gene expression analysis using RT-PCR. Canadian Journal of Microbiology. 2016;**62**(5):431-436. DOI: 10.1139/cjm-2015-0734

[40] Zhao S, Yin R, Zhang M, Zhai Z, Shen Z, Mou Y, et al. Efficient gene editing in the slow-growing, non-sporulating, melanized, endophytic fungus *Berkleasmium* sp. Dzf12 using a CRISPR/Cas9 system. World Journal of Microbiology and Biotechnology. 2024;**40**(6):1-13. DOI: 10.1007/s11274-024-03988-y

[41] Yang Q, Wu M, Zhu YL, Yang YQ, Mei YZ, Dai CC. The disruption of the MAPKK gene triggering the synthesis of flavonoids in endophytic fungus *Phomopsis liquidambaris*. Biotechnology Letters. 2020;**43**(1):119-132. DOI: 10.1007/s10529-020-03042-5

[42] Kang X, Liu C, Shen P, Hu L, Lin R, Ling J, et al. Genomic characterization provides new insights into the biosynthesis of the secondary metabolite Huperzine a in the endophyte *Colletotrichum gloeosporioides* Cg01. Frontiers in Microbiology. 2019;**9**:3237. DOI: 10.3389/fmicb.2018.03237

[43] Paulo B, Sigrist R, Angolini C, Eberlin M, de Oliveira L. Gene deletion leads to improved Valinomycin production by *Streptomyces* sp. CBMAI 2042. Journal of the Brazilian Chemical Society. 2018;**30**(3):673-679. DOI: 10.21577/0103-5053.20180207

[44] Fidan O, Zhan J. Discovery and engineering of an endophytic *pseudomonas* strain from *Taxus chinensis* for efficient production of zeaxanthin diglucoside. Journal of Biological Engineering. 2019;**13**:66. DOI: 10.1186/s13036-019-0196-x

[45] Bian G, Yuan Y, Tao H, Shi X, Zhong X, Han Y, et al. Production of taxadiene by engineering of mevalonate pathway in *Escherichia coli* and endophytic fungus Alternaria alternata TPF6. Biotechnology Journal. 2017;**12**(4):1600697. DOI: 10.1002/biot.201600697

[46] Wang M, Liu S, Li Y, Xu R, Lu C, Shen Y. Protoplast mutation and genome shuffling induce the Endophytic fungus *Tubercularia* sp. TF5 to produce new compounds. Current Microbiology. 2010;**61**(4):254-260. DOI: 10.1007/s00284-010-9604-7

[47] Zhao K, Zhou DP, Ping WX, Ge JP. Study on the preparation and regeneration of protoplast from taxol-producing fungus *Nodulisporium sylviforme*. Nature Science. 2004;**2**:52-59

[48] Xuanwei Z, Yamin W, Huifang Z, Zinan W, Juan L, Lu L, et al. Protoplast formation, regeneration and transformation from the taxol-producing

fungus *Ozonium* sp. African Journal of Biotechnology. 2008;7(12):2017-2024. DOI: 10.5897/ajb2008.000-5050

[49] Kai Z, Sun Q, Zhang Y, Ping W, et al. Screening and characterization of a high taxol producing fungus by protoplast mutagenesis. High Technology Letters. 2009;**15**(2):220-226

[50] Xin Y, Yang M, Yin H, Yang J. Improvement of ethanol tolerance by inactive protoplast fusion in *Saccharomyces cerevisiae*. BioMed Research International. 2020;**2020**:1-10. DOI: 10.1155/2020/1979318

[51] Gnanam AJ. Protoplast fusion techniques in fungi. In: Gupta VK, Tuohy MG, editors. Laboratory Protocols in Fungal Biology. New York, NY: Springer; 2012. pp. 483-488. DOI: 10.1007/978-1-4614-2356-0_45

[52] Zhao K, Zhou DP, Ping WX, Liu J, Ma X, Jin T. Study on breeding up high-yield strain of taxol by protoplast mutagensis. PubMed. 2005;**21**(5):848-851

[53] Zhou J, Wang B, Zhu L. Conditioned culture for protoplasts isolated from *chrysanthemum*: An efficient approach. Colloids and Surfaces B: Biointerfaces. 2005;**45**(3-4):113 119. DOI: 10.1016/j.colsurfb.2005.07.012

[54] Zhang Y, Su J, Duan S, Ao Y, Dai J, Liu J, et al. A highly efficient rice green tissue protoplast system for transient gene expression and studying light/chloroplast-related processes. Plant Methods. 2011;7:30. DOI: 10.1186/1746-4811-7-30

[55] Arnau J, Yaver D, Hjort CM. Strategies and challenges for the development of industrial enzymes using fungal cell factories. In: Nevalainen H, editor. Grand Challenges in Fungal Biotechnology. Cham, Switzerland: Springer; 2020. pp. 179-210. DOI: 10.1007/978-3-030-29541-7_7

[56] Li D, Tang Y, Lin J, Cai W. Methods for genetic transformation of filamentous fungi. Microbial Cell Factories. 2017;**16**:168. DOI: 10.1186/s12934-017-0785-7

[57] Verma V, Batta A, Singh HB, Srivastava A, Garg SK, Singh VP, et al. Bioengineering of fungal endophytes through the CRISPR/Cas9 system. Frontiers in Microbiology. 2023;**14**:1146650. DOI: 10.3389/fmicb.2023.1146650

[58] Lichius A, Ruiz DM, Zeilinger S. Genetic transformation of filamentous fungi: Achievements and challenges. In: Nevalainen H, editor. Grand Challenges in Fungal Biotechnology. Cham, Switzerland: Springer; 2020. pp. 123-164. DOI: 10.1007/978-3-030-29541-7_5

[59] Balabanova LA, Shkryl YN, Slepchenko LV, Yugay YA, Gorpenchenko TY, Kirichuk NN, et al. Development of host strains and vector system for an efficient genetic transformation of filamentous fungi. Plasmid. 2019;**101**:1-9. DOI: 10.1016/j.plasmid.2018.11.002

[60] Nevalainen KMH, VSJ T'o, Bergquist PL. Heterologous protein expression in filamentous fungi. Trends in Biotechnology. 2005;**23**(9):468-474. DOI: 10.1016/j.tibtech.2005.06.002

[61] Hur JY, Jeong E, Kim YC, Lee SR. Strategies for natural product discovery by unlocking cryptic biosynthetic gene clusters in fungi. Separations. 2023;**10**(6):333. DOI: 10.3390/separations10060333

[62] Yamin W, Xuanwei Z, Lu L, Jie L, Zinan W, Geng Y, et al. An efficient transformation system of

taxol-producing endophytic fungus EFY-21 (*Ozonium* sp.). African Journal of Biotechnology. 2010;**9**(12):1726-1733. DOI: 10.5897/ajb2010.000-3019

[63] Du Y, Liu Y, Hu J, Peng X, Liu Z. CRISPR/Cas9 systems: Delivery technologies and biomedical applications. Asian Journal of Pharmaceutical Sciences. 2023;**18**(6):100854. DOI: 10.1016/j.ajps.2023.100854

[64] Song R, Zhai Q, Sun L, Huang E, Zhang Y, Zhu Y, et al. CRISPR/Cas9 genome editing technology in filamentous fungi: Progress and perspective. Applied Microbiology and Biotechnology. 2019;**103**(17):6919-6932. DOI: 10.1007/s00253-019-10007-w

[65] Schuster M, Kahmann R. CRISPR-Cas9 genome editing approaches in filamentous fungi and oomycetes. Fungal Genetics and Biology. 2019;**130**:43-53. DOI: 10.1016/j.fgb.2019.04.016

[66] Salazar-Cerezo S, Kun RS, de Vries RP, Garrigues S. CRISPR/Cas9 technology enables the development of the filamentous ascomycete fungus *Penicillium subrubescens* as a new industrial enzyme producer. Enzyme and Microbial Technology. 2020;**133**:109463. DOI: 10.1016/j.enzmictec.2019.109463

[67] Satish L, Shamili S, Muthubharathi BC, Ceasar SA, Kushmaro A, Singh V, et al. CRISPR-Cas9 system for fungi genome engineering toward industrial applications. In: Singh V, Dhar PK, editors. Genome Engineering via CRISPR-Cas9 System. Amsterdam, Netherlands: Elsevier Science; 2020. pp. 69-81. DOI: 10.1016/b978-0-12-818140-9.00006-4

[68] Zhou J, Li X, Huang PW, Dai CC. Endophytism or saprophytism: Decoding the lifestyle transition of the generalist fungus *Phomopsis liquidambari*. Microbiological Research. 2018;**206**:99-112. DOI: 10.1016/j.micres.2017.10.005

[69] Salazar-Cerezo S, de Vries RP, Garrigues S. Strategies for the development of industrial fungal producing strains. Journal of Fungi. 2023;**9**(8):834. DOI: 10.3390/jof9080834

[70] Zhu J, Yan L, Xu X, Zhang Y, Shi J, Jiang C, et al. Strategies to enhance the production of pinoresinol and its glucosides by endophytic fungus (*Phomopsis* sp. XP-8) isolated from Tu-chung bark. AMB Express. 2018;**8**:55. DOI: 10.1186/s13568-018-0584-5

[71] Magotra A, Kumar M, Kushwaha M, Awasthi P, Raina C, Gupta AP, et al. Epigenetic modifier induced enhancement of fumiquinazoline C production in *aspergillus fumigatus* (GA-L7): An endophytic fungus from *Grewia asiatica* L. AMB Express. 2017;**7**:43. DOI: 10.1186/s13568-017-0343-z

[72] Varasteh-kojourian M, Ganjeali A, Asili J, Malekzadeh-Shafaroudi S, Taleghani A. 5-azacytidine increases Tanshinone production in Endophytic fungi of *Salvia abrotanoides* (Kar.) *Sytsma*. Journal of Cell and Molecular Research. 2023;**14**(2):78-84

[73] Wakefield J, Hassan HM, Jaspars M, Ebel R, Rateb ME. Dual induction of new microbial secondary metabolites by fungal bacterial co-cultivation. Frontiers in Microbiology. 2017;**8**:1284. DOI: 10.3389/fmicb.2017.01284

[74] Akone SH, Pham CD, Chen H, Ola ARB, Ntie-Kang F, Proksch P. Epigenetic modification, co-culture and genomic methods for natural product discovery. Physical Sciences Reviews. 2018;**4**(4):20180118. DOI: 10.1515/psr-2018-0118

[75] Peng XY, Wu JT, Shao CL, Li ZY, Chen M, Wang CY. Co-culture: Stimulate the metabolic potential and explore the molecular diversity of natural products from microorganisms. Marine Life Science & Technology. 2021;**3**(3):363-374. DOI: 10.1007/s42995-020-00077-5

[76] Gupta S, Kulkarni MG, White JF, Van Staden J. Epigenetic-based developments in the field of plant endophytic fungi. South African Journal of Botany. 2020;**134**:394-400. DOI: 10.1016/j.sajb.2020.07.019

[77] Cichewicz RH. Epigenome manipulation as a pathway to new natural product scaffolds and their congeners. Natural Product Reports. 2010;**27**(1):11-22. DOI: 10.1039/b920860g

[78] Belofsky GN, Anguera M, Jensen PR, Fenical W, Köck M. Oxepinamides A-C and Fumiquinazolines H-I: Bioactive metabolites from a marine isolate of a fungus of the genus *Acremonium*. Chemistry—A European Journal. 2000;**6**(8):1355-1360. DOI: 10.1002/(sici)1521-3765(20000417)6:8<1355::aid-chem1355>3.3.co;2-j

Section 2

# Antioxidant Potential of Phytochemicals

Chapter 3

# Antioxidant and Prooxidant Functions of Carotenoids in Human Health: Trigger Factors, Mechanism and Application

*Aisha Farhana, Yusuf Saleem Khan, Abdullah Alsrhani, Emad Manni, Ayman A.M. Alameen, Wassila Derafa, Nada Alhathlaul, Muhammad Atif and Lienda Bashier Eltayeb*

## Abstract

Carotenoids are plant-derived compounds that have numerous health benefits, encompassing disease protection, cardiovascular health, mental and physical development, etc. Carotenoids exert their effect by combating oxidative stress, scavenging free radicals and quenching singlet oxidants, hence an effective metabolic antioxidant. Recently, carotenoids have gained a debated interest, wherein research momentum diverges to understand their antioxidant as well as prooxidant functions. Research focuses on investigating the prooxidant function stemming from the ability of carotenoids to induce oxidative damage. Herein, we will discuss the general classification, structure and physiological reactions of carotenoids. The emphasis will lie in understanding the triggers and mechanisms that influence their antioxidant versus prooxidant ability. We aim to outline the likely conditions enabling their antioxidant and/or prooxidant activities. A thorough mechanistic understanding of these highly beneficial nutrients will provide a toolkit for preventing diseases in addition to treating diseases through their exclusive prooxidant abilities.

**Keywords:** carotenoids, plant-derived compounds, health benefits, antioxidant mechanism, antioxidant-prooxidant balance, oxidative stress, prooxidant, free radical scavenging, singlet oxygen quenching, metabolic antioxidant

## 1. Introduction

Carotenoids are a class of over 700 fat-soluble pigments, primarily synthesized in plants, algae and certain bacteria and found in some animal products [1]. These lipophilic isoprenoid compounds contribute to the vibrant colors in plants and vegetables, ranging from yellow, orange to red hues [2]. About 20 carotenoids have been found in human blood and tissues, while about 30–40 carotenoids have been identified in the average human diet [3]. Carotenoids play various important roles in plant and animal physiology with significant benefits to human health due to their antioxidant properties [4–6]. This makes carotenoids the key bioactive molecules in various fields, including nutrition, medicine and biotechnology [7].

Carotenoids are usually thought of as antioxidants because of their ability to neutralize free radicals and lower oxidative stress, but in certain incidences, they may also behave as prooxidants [8, 9]. A number of variables, including chemical composition and concentration of the carotenoid, metal concentration, light exposure, the impact of other biological antioxidants and the intrinsic molecular structure of carotenoids, might influence this functional alteration [8, 10, 11]. Therefore, evaluating their impacts on health underlies the critical assessment of the overall context, which includes dose, environment and the presence of other chemicals [12]. They play essential roles in photosynthesis and photoprotection in plants while offering significant health benefits to humans. Carotenoids are broadly categorized into two classes: carotenes and xanthophylls [8]. Carotenes, such as β-carotene and lycopene, are purely hydrocarbon compounds, whereas xanthophylls, like lutein and zeaxanthin, contain oxygen [12]. These structural variations contribute to their diverse biological functions. In humans, certain carotenoids act as precursors to vitamin A, a crucial nutrient for vision and immune function [13]. They also exhibit potent antioxidant properties, neutralizing reactive oxygen and nitrogen species (ROS/RNS) and mitigating oxidative stress connected to chronic illnesses and aging [14, 15].

Carotenoids encompass major health benefits to human health [16]. Beyond their antioxidant properties, the bioactivity spectrum of carotenoids extends to its potential as prooxidant. Emerging research highlights the potential of carotenoids to reduce the risk of developing chronic illnesses including heart disease, age-related macular degeneration (AMD) and several types of cancer. Some important roles of carotenoids include improving cellular defense mechanisms, regression of malignancy, growth factor and inflammatory pathways modulation. Nonetheless, the effect underlies context-dependence, wherein, under certain conditions, carotenoids switch to prooxidant functions, especially when present in high concentrations or exposed to high levels of oxidative stress, such as in smokers. The scientific community is still debating on safe and effective dietary intake or supplementation of carotenoids because of their dual nature, which highlights the intricacy of their physiological activities. This chapter explores the dual roles of carotenoids, detailing their mechanisms as antioxidants and prooxidants, delving into the intricate dual roles of carotenoids and highlighting their mechanisms of action as both potent antioxidants and, under certain conditions, prooxidants.

## 2. Structural and biochemical foundations of carotenoids

The characteristic hues of carotenoids, which vary from yellow and orange to red, are attributed to a sequence of conjugated double bonds [1]. Within the molecule, these double bonds are usually organized either cyclically or linearly [17]. Since light in the visible spectrum is absorbed by the conjugated system of double bonds, carotenoids are crucial pigments of plants, algae and some microbes [18]. A long sequence of conjugated double bonds, usually organized in a polyene chain, is the primary structural characteristic of carotenoids. The π-electron cloud is extended along the molecule by this conjugation, which causes visible light to be absorbed and gives carotenoids their distinctive coloring [19]. Nonetheless, their electron-rich system makes them a subject of electrophilic attack, thereby imparting greater instability toward oxidation.

## 2.1 Molecular diversity and functional groups

Based on their chemical structures and functional groups, carotenoids are segregated into two main groups: carotenes and xanthophylls (**Figure 1**). As hydrocarbons, carotenes are devoid of functional groups with oxygen, solely harboring carbon and hydrogen atoms. This property differentiates carotenes from xanthophylls and also makes them lipophilic. Further classification differentiates carotenoids to be pro-vitamin A (α-carotene, β-carotene and β-cryptoxanthin) and non-pro-vitamin A (lutein, lycopene and zeaxanthin), depending on their ability or inability as precursors of vitamin A (**Figure 1**).

Typical carotenoid examples include β-carotene, α-carotene and lycopene.

*β-carotene*: It is an efficient electron delocalization tetraterpene made up of 40 carbon atoms with 11 conjugated double bonds (**Figure 1**). The vivid orange hue of the compounds results from the absorption of visible light by its expanded π-electron system [20]. Its antioxidant qualities are also influenced by this delocalized electron network, efficiently neutralizing free radicals and quenching singlet oxygen [21]. β-Carotenes are present in sweet potatoes, pumpkins and carrots and are a precursor of vitamin A.

*α-carotene:* The structure of α-carotene is similar to that of β-carotene, except that it lacks one β-ionone ring [22]. It gives green leafy vegetables their yellow-orange hue and is found in large quantities.

*Lycopene:* Lycopene is a strong scavenger of free radicals due to its open chain polyene structure and absence of β-ionone rings, delineating them from other carotenoids like β-carotene. The linear configuration of conjugated double bonds improves its capacity to neutralize ROS and quench singlet oxygen [23]. Tomatoes, watermelons and pink grapefruits are rich sources of the red pigment lycopene.

Xanthophylls are often classified as lutein, zeaxanthin, β-cryptoxanthin, canthaxanthin and astaxanthin.

Being the derivatives of oxidized carotenes, xanthophylls have conjugated double bonds together with oxygen-containing functional groups like hydroxyl (−OH) or epoxy (−O−) groups [24]. Oxygen atoms provide xanthophylls with special chemical

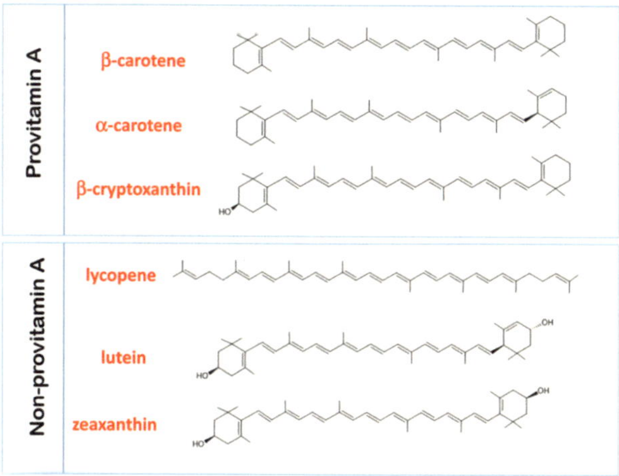


**Figure 1.**
*Molecular structure of carotenoids. (a) Carotenes and (b) Xanthophylls. The efficacy of carotenoids as antioxidants dependent on the ability of these molecules to bind at the water/lipid interface.*

characteristics and biological roles [17]. It makes up 10% of the carotenoids found in nature [25]. Typical xanthophylls include:

*Lutein:* Eight isoprene units make up the 40-carbon backbone of lutein, an unsaturated polyene hydrocarbon. Its unique structure and functional characteristics are attributed to the two hydroxyl groups that are joined to β-ionone rings [17, 26]. Lutein, which is concentrated in the macula of the retina and is necessary for eye health, is available from egg yolks, marigold flowers and green leafy foods.

*β-cryptoxanthin:* It contains 11 conjugated double bonds similar to polyisoprenoids (**Figure 1**). Structurally, it is a bicyclic compound with a hydroxyl group attached to the carbon at position 3 of one of its rings [20]. Retinol, or vitamin A, is produced by the metabolic conversion of β-cryptoxanthin in the human body and is essential for both eyesight and general health.

*Zeaxanthin:* Nine alternating conjugated double and single carbon bonds make up this extremely reactive and stable polyene molecule [15]. Zeaxanthins have hydroxyl-substituted ionone rings that cap each end of its carbon skeleton enhancing its solubility and biological use [26]. Because of its structure, zeaxanthin can efficiently interact with ROS, which helps to protect cells from oxidative damage and preserve eye health, especially by blocking damaging blue light in the macula [27]. Zeaxanthin is often found in orange peppers, spinach and maize. It is considered an essential carotenoid for maintaining vision and A.

*Astaxanthin:* This xanthophyll is made up of two terminal rings of the β-ionone type that are joined by a polyene chain. This ketocarotinoids contain β-ionone ring with hydroxyl group (−OH) at both ends of the molecule and an asymmetric carbon at the 3,3′-position [28]. The ring structure also has oxygen as hydroxyl (−OH) and keto groups (C=O) [29]. Its strong antioxidant qualities and functions in the biological systems are facilitated by special structural arrangement [30]. Astaxanthin is a red pigment found in seafood, such as lobster, shrimp and salmon, flamingos and crustaceans that gives them pink color.

## 2.2 Physiological relevance of carotenoids in human metabolism

The chemical and physical properties of carotenoids dictate their functional role in human metabolism, serving both as antioxidants and precursors to vital biomolecules (**Figure 2**). β-Carotene, for instance, is converted into vitamin A, which

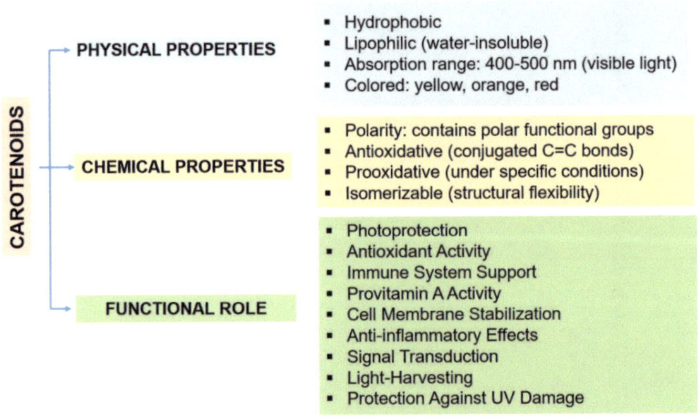


**Figure 2.**
*Overview of physical and chemical properties of carotenoids and functional roles.*

is critical for vision, immune function and cellular differentiation. Carotenoids, lutein and zeaxanthin are concentrated in the retina, particularly the macula, where they prevent oxidative damage, thus supporting eye health and reducing the risk of age-related macular degeneration (AMD) [31]. Additionally, carotenoids have been linked to improved cardiovascular health by reducing lipid oxidation and enhancing endothelial function [32].

Carotenoids undergo several key biochemical pathways in human metabolism. Once ingested, carotenoids are absorbed in the intestine, aided by dietary fats and bile salts and incorporated into chylomicrons for transport *via* the lymphatic system. β-Carotene is enzymatically cleaved by β-carotene-15,15′-dioxygenase in the intestine to yield retinal, which can be reduced to retinol (vitamin A) or oxidized to retinoic acid [33]. These metabolites are indispensable for vision, epithelial cell differentiation and immune regulation, and their deficiencies are linked to conditions like xerophthalmia and increased susceptibility to infections [34–36]. Non-provitamin A carotenoids, such as lutein and zeaxanthin are concentrated in the macula and create the macular pigment optical density (MPOD) [36]. They reduce oxidative stress and function as blue light filters, which lowers the risk of AMD [36].

Carotenoids play a pivotal role in modulating critical signal transduction pathways that regulate cellular responses to oxidative stress, inflammation and growth regulation. Key pathways influenced by carotenoids include nuclear factor erythroid 2-related factor 2 (Nrf2) and nuclear factor-kappa B (NF-κB) [37]. Through Nrf2 activation, carotenoids, such as lycopene and astaxanthin enhance the expression of antioxidant and detoxifying enzymes, strengthening the defense against oxidative damage. Simultaneously, carotenoids inhibit NF-κB signaling, which reduces the expression of pro-inflammatory cytokines and mitigates chronic inflammation [38]. NF-κB also governs genes associated with embryonic development, skin and bone health and the central nervous system. Specific interactions of carotenoids with signaling pathways such as PI3K/Akt, Wnt/β-catenin, MAPK, JAK/STAT, mTOR, TGF-β and Hedgehog have been well-studied [39–42]. These pathways are pivotal in regulating cellular stress, growth, metabolism, embryonic development, immune responses, inflammation and the inhibition of cancerous proliferation. By modulating these cascades, carotenoids exhibit remarkable therapeutic potential in the prevention and management of chronic diseases, including cancer, cardiovascular disorders and neurodegenerative conditions, thereby maintaining metabolic homeostasis.

Carotenoids also have the potential to activate nuclear hormone receptor pathways such as retinoic acid receptors (RAR), retinoid X receptors (RXR) and peroxisome proliferator-activated receptors (PPARs) [43, 44]. Both *in vitro* and animal studies have connected Nrf-2, a transcription factor that activates antioxidant response elements, to carotenoid-mediated oxidative stress reduction [45, 46]. When activated, Nrf2 upregulates the expression of antioxidant enzymes, including glutathione peroxidase, SOD and catalase [47]. These enzymes work synergistically with carotenoids to reduce oxidative stress, mitigate inflammation and protect cellular structures, including lipids, proteins and DNA, from oxidative damage. Carotenoids also play important roles in apoptosis, cellular proliferation, metabolic inflammation and tumor growth. The RAR and RXR receptors also regulate the expression of several genes associated with immunological function and cellular development [48]. Furthermore, RXR/PPARs have been linked to adipocyte differentiation [49]. Studies have also shown that carotenoids activate PPARγ and Nrf-2 [47].

Additionally, carotenoids contribute to cellular communication by enhancing gap junctional intercellular communication (GJIC), a process crucial for maintaining tissue homeostasis and regulating cell growth [50]. It has been particularly demonstrated that carotenoids like β-carotene increase the expression of connexin proteins, which are the structural elements of gap junctions [50]. GJIC is strengthened by this regulatory mechanism, which can subsequently stalls tumor growth through unregulated cell division [51]. The capacity of carotenoids to repair and enhance GJIC underlies their function in preventing cancer and their wider significance in preserving cellular integrity.

### 2.3 Redox potential and reactivity: The basis of dual functionality as antioxidant and prooxidant

The dual roles of carotenoids as antioxidants and prooxidants are dictated by their redox potential and chemical reactivity (**Figure 2**). The ability to donate or take electrons in redox processes is a result of the redox potential, which is inherent in their conjugated structure of double bonds [52]. Carotenoids are antioxidants that stabilize free radicals and efficiently quench singlet oxygen by donating electrons or hydrogen atoms; this antioxidant effect is more obvious during mild oxidative stress [41]. Their redox potential is usually within a range conducive of scavenging ROS, which makes them effective at maintaining cellular redox homeostasis. Carotenoids may, however, behave as prooxidants in some instances, adding to oxidative stress. Carotenoid-derived radicals can emerge when the antioxidant action of carotenoids is destabilized by high concentrations of carotenoids, low or high oxygen tension and environments with enhanced ROS [52]. These radicals worsen cellular oxidative damage by interacting with cellular macromolecules and spreading lipid peroxidation. Additionally, the presence of transition metals like iron or copper and prolonged exposure to UV or blue light, can change the redox state of carotenoids, shifting the balance in favor of prooxidant activity [53]. The delicate balance between prooxidant and antioxidant functions is further highlighted by the fact that their oxidative metabolites and coexistence with pre-existing lipid peroxides can enhance prooxidant effects. This dual functionality emphasizes the significance of context such as concentration, oxygen availability and degrees of oxidative stress in deciding whether carotenoids have a protective or detrimental effect on biological systems.

#### *2.3.1 Antioxidant effects of carotenoids: Guardians against oxidative stress*

Carotenoids are powerful antioxidants, acting through mechanisms like scavenging reactive species, electron transfer, hydrogen abstraction, radical addition, metal chelation, quenching molecular oxygen and preventing lipid peroxidation. Their chemical composition, interactions with biological membranes and co-antioxidants like vitamins C and E influence their antioxidant activity. The quantity of conjugated double bonds and the type of substituents in the carotenoids, which include cyclic end groups, also influence the quenching activity.

Astaxanthin is ten times more powerful than lutein, canthaxanthin, and β-carotene. It helps stop lipid peroxidation in cellular membranes. [54]. Astaxanthin has scavenging capabilities against ROA, including superoxide anion ($O\text{-}_2$) and hydrogen peroxide ($H_2O_2$) radicals. The natural astaxanthin found in shrimp shells has antioxidant properties that protect against non-physiological radicals [55]. Astaxanthin lowers ROS formation and reduces ROS in murine cone cell lines in a concentration-dependent manner. It boosts the activity of antioxidant enzymes and decreases MDA levels in human LS-180 colorectal cancer cells. *In vivo*, it protects the heart from oxidative stress, promotes anti-arthritic effects in rats and provides neuroprotection in the mouse brain [56, 57]. In HR-1 hairless mice, astaxanthin also prevents photoaging, demonstrating its wide-ranging protective benefits in a variety of models that have been investigated thus far [58]. Astaxanthin supplementation was demonstrated to dramatically lower oxidative stress indicators, such as MDA and ROS, in healthy people in clinical research [59]. Further, consumption of astaxanthin improved skin elasticity and reduced the effects of photoaging in those exposed to UV radiation [60]. These findings lend credence to broad-spectrum antioxidant activity of astaxanthin and its capacity to lessen oxidative damage in humans.

Alpha-carotene has shown efficacy in scavenging free radicals, thereby reducing lipid peroxidation in cultured human cells [61]. Supplementation with α-carotene dramatically reduced MDA, a hallmark of oxidative damage, subsequently increasing the activity of antioxidant enzymes including SOD and catalase. Many studies also demonstrate its function in cellular defense. In a clinical trial of people with metabolic syndrome, α-carotene supplementation was found to dramatically lower oxidative stress indicators, such as MDA and ROS metabolites [62]. Higher dietary consumption of α-carotene was linked to decreased oxidative stress and reduced risk of cardiovascular diseases in a Japanese people cohort, according to a study by Hozawa et al. [63]. These studies underscore the beneficial potential of α-carotene in reducing oxidative damage in humans.

β-Carotene is a precursor of vitamin A and illustrates the ability to scavenge free radicals like ABTS· + and ROS such as $^1O_2$, ONOO, peroxynitrous acid (ONOOH) and $NO_2$. Its scavenging activity underlies its location and orientation in biological membranes and isomerization states. The lipophilic nature of β-carotene facilitates its efficacy in the lipidic systems. β-Carotene supplementation is shown to decrease oxidative stress in smokers by lowering lipid peroxidation, confirming its protective function against oxidative damage [64]. β-Carotene is shown to increase skin elasticity and decrease UV-induced oxidative damage in humans [65]. This property is further supported by numerous experimental investigations. The capacity of β-carotene to prevent lipid peroxidation in liposomal membranes is dependent on its orientation and location inside the membrane, affecting its radical scavenging capacity [66]. Furthermore, β-carotene confers protection on vascular health by dramatically lowering ROS levels in human endothelial cells [67].

Lutein can scavenge ROS such as ABTS· +, NO· and $O_2$·. *In vitro* studies demonstrate the inhibition of lipid peroxidation and scavenging of DPPH radicals [68]. Lutein supplementation in humans enhances antioxidant enzyme system and reduces oxidative stress, particularly in the liver and blood [69]. Lutein also offers protective effects against oxidative damage-associated diseases such as ovarian ischemia-reperfusion injury and cyclophosphamide-induced lung damage [70]. Several *in vitro* studies have demonstrated that lutein readily neutralizes the ABTS· + radical and subdues oxidative stress markers in biological systems. In animal models, lutein has been shown to enhance catalase and SOD activities, while reducing oxidative damage in liver and lungs. In rats with ovariectomies, lutein supplementation raised blood glutathione levels and decreased lipid peroxidation [71]. Lutein supplementation enhances cognitive performance in older persons, especially in domains like memory and attention. fMRI studies have demonstrated that lutein has neuroprotective benefits *via* increasing brain activity and connection [72].

Lycopene is extremely effective in neutralizing ROS, that is, $^1O_2$, $O_2$· and HO·. It demonstrates greater effectiveness in preventing oxidative damage and scavenging ROS. Lycopene demonstrates greater capacity to quench ROS, reduce ROS formation driven by light exposure in murine cone cell lines, prevent hemoglobin oxidation and lipid peroxidation in human erythrocytes [73]. Additionally, studies show that lycopene protects rats exposed to hepatotoxic chemicals such as aflatoxin B1 and bisphenol A by lowering ROS and MDA levels and increasing total antioxidant capacity [74, 75]. In humans lycopene intake lowers oxidative stress, especially in cardiovascular and neurological disease.

Zeaxanthin, plentiful in dietary and marine sources, is essential for controlling the production of glutathione, enhancing redox equilibrium and reducing cellular damage brought on by $H_2O_2$. Zeaxanthin has strong action against hydroxyl radicals (HO•) and effectively scavenges ABTS$^{\cdot}$ + and quenches singlet oxygen ($^1O_2$) [76]. Additionally, it interacts with oxidized LDL and shields the probe DHR 123 from $ONOO^-$ oxidation. Zeaxanthin is less efficient against tert-butyl hydroperoxide-induced lipid damage in human erythrocytes, but it does decrease APPH-induced lipid peroxidation and hemoglobin oxidation. Reduced MDA levels and increased activity of antioxidant enzymes such as glutathione, SOD and catalase are indicators of its protective benefits in acetic acid-induced colitis in rats, according to *in vivo* studies [77]. Zeaxanthin consumption may help prevent AMD and other oxidative stress-related disorders by lowering oxidative stress and concentrating in the macula, according to human research.

Hypoxic conditions reduce ROS production, thereby minimizing oxidative stress and aiding in cellular protection. Carotenoids such as β-carotene, lutein and lycopene display potent antioxidant activity by scavenging free radicals without transitioning into prooxidant states, more likely at higher oxygen levels. This function is critical for maintaining cellular homeostasis, particularly in oxygen-sensitive tissues like the retina. In this case, lutein and zeaxanthin, which are mostly found in the macula, protect against phototoxic damage by blocking ROS and lipid peroxidation. This keeps the metabolism in balance and keeps cells healthy in oxidative environments.

### *2.3.2 Mechanisms underpinning the antioxidant activity of carotenoids*

Carotenoids interact with free radicals such as superoxide anions ($O_2^-$ hydroxyl radicals (HO·) and nitrogen-derived radicals, neutralizing them by electron transfer or by donating hydrogen atoms. For example, β-carotene scavenges the non-physiological ABTS radical cation (ABTS•+) and neutralizes peroxynitrous acid (ONOOH), an oxidizing agent implicated in cellular damage.

Low oxygen tension and balanced dietary intake lead to the optimal antioxidant activity of carotenoids. Within physiological levels, carotenoids mitigate oxidative stress and inflammation, contributing to cellular redox homeostasis. However, excessive carotenoid concentration, especially in absence of any complementary antioxidant nutrients induces prooxidant effects that tips off the redox balance thereby worsening oxidative damage. Maintaining appropriate carotenoid levels is critical for preserving cellular integrity, preventing oxidative damage and sustaining metabolic health. Carotenoids carry out antioxidant mechanisms in the following ways:

#### *2.3.2.1 Direct free radical scavenging*

Carotenoids interact with free radicals such as superoxide anions ($O_2^-$), hydroxyl radicals (HO· and nitrogen-derived radicals, neutralizing them by electron transfer or by donating hydrogen atoms. As an example, β-carotene gets rid of the non-physiological ABTS radical cation (ABTS$^{\cdot}$+) and neutralizes peroxynitrous acid

(ONOOH), which is an oxidizing agent that is linked to cell damage. Lycopene, due to its extended conjugation, exhibits higher efficiency in quenching free radicals like hydroxyl radicals compared to other carotenoids [78].

#### *2.3.2.2 Singlet oxygen quenching*

Singlet oxygen ($O_2$) is a very reactive type of oxygen that may seriously harm cells, inflicting oxidative damage. Through a physical process called energy transfer, the energy from singlet oxygen is transferred to the carotenoid molecule, quenching singlet oxygen [79]. By releasing the absorbed energy as heat, the carotenoid returns to its ground state and the singlet oxygen is changed into its less reactive triplet state [79]. Carotenoids, lycopene and astaxanthin are essential for avoiding lipid peroxidation and safeguarding cellular membranes because of their long chains of conjugated double bonds, which make them extremely effective quenchers of singlet oxygen. Carotenoids, like lutein or astaxanthin, neutralize the damaging reactivity of singlet oxygen by absorbing its excess energy and dissipating it as heat when they come into contact with it [5]. This method shields cellular structures against oxidative damage, especially in high-exposure organs like the retina, by preventing singlet oxygen from interacting with adjacent molecules like lipids and proteins.

#### *2.3.2.3 Interaction with cellular antioxidant system*

Carotenoids function synergistically with intrinsic antioxidant mechanisms. They augment the function of enzymes including SOD, catalase and glutathione peroxidase by diminishing ROS levels and preserving intracellular redox equilibrium [42]. This not only mitigates oxidative stress but also maintains the efficacy of other antioxidants such as glutathione. This emphasize the essential function of carotenoids in safeguarding against oxidative damage, preserving cellular integrity and alleviating chronic illnesses such as cardiovascular diseases, neurological disorders and certain malignancies (**Table 1**).

Further, carotenoids exhibit antioxidant action through direct and indirect mechanisms. Directly, they scavenge ROS and lipid radicals, mitigating oxidative stress (**Figure 3**). Indirectly, they influence cellular signaling pathways related to redox homeostasis, supporting the antioxidant defense system [80]. However, under certain conditions, carotenoids can also behave as prooxidants by interacting with oxygen or lipid radicals, amplifying the lipid peroxidation chain reaction (**Table 1**) [8].

The indirect antioxidant action of carotenoids involves their metabolites, such as electrophilic apocarotenoids, which are cleavage products of parent carotenoids by β-carotene oxygenase 2. These apocarotenoids activate Nrf2, *via* specific signal transduction pathways [81]. Additionally, ketocarotenoids-oxidative metabolites of xanthophylls with α,β-unsaturated carbonyl structures – exhibit electrophilic properties that may also contribute to this process. The efficacy of carotenoids as biological antioxidants is influenced by their cellular and extracellular distribution, underscoring the importance of their bioavailability and precise interactions in mitigating oxidative stress.

### 2.4 Prooxidant activities of carotenoids: When protection turns to peril

Carotenoids though known for their antioxidant properties, can paradoxically exhibit prooxidant activities under specific conditions [8]. As discussed earlier, these effects are highly context-dependent [8]. The mechanism by which carotenoids

| Function | Mechanism | Trigger factors | Examples of carotenoids | Examples of applications |
|---|---|---|---|---|
| Antioxidant | Neutralization of free radicals by donating electrons | Adequate oxygen levels, low ROS concentration | Beta-carotene, lutein, zeaxanthin | Prevention of oxidative stress-related diseases cardiovascular diseases, cancer prevention |
| Prooxidant | Generation of ROS under high concentrations or imbalance | High oxygen tension, excessive carotenoid intake, metal interaction | Beta-carotene, lycopene | Potential use in targeted cancer therapies (inducing oxidative stress in cancer cells) |
| Synergistic function | Interaction with other antioxidants (vitamin E, C) to enhance effects | Balanced diet with complementary nutrients | Astaxanthin, lutein | Dietary supplements and functional foods |
| Regulatory role | Modulation of gene expression pathways linked to redox balance | Genetic predisposition, environmental factors | Lycopene, beta-cryptoxanthin | Development of nutraceuticals for personalized medicine |

**Table 1.**
*Dual role of carotenoids in human health and influencing factors.*

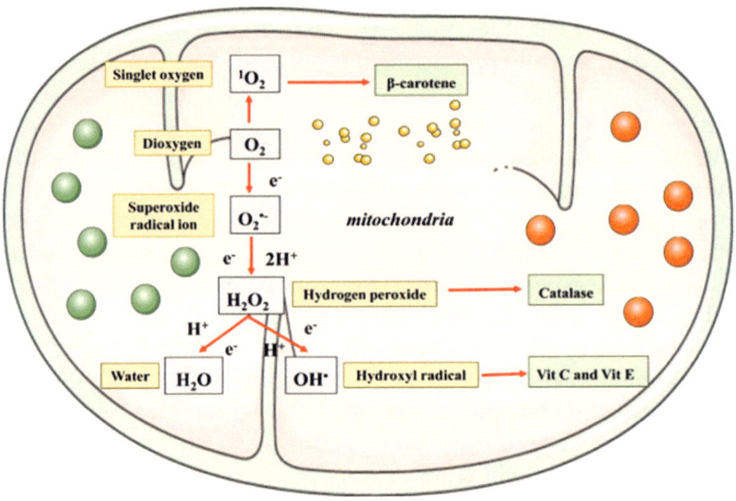


**Figure 3.**
*The diagram depicts the production and elimination of reactive oxygen species (ROS) in mitochondria, emphasizing the oxidative stress pathway. Molecular oxygen ($O_2$) participates in successive redox processes, yielding superoxide radicals ($O_2^{\cdot-}$), hydrogen peroxide ($H_2O_2$) and hydroxyl radicals ($OH\cdot$), whereas singlet oxygen ($^1O_2$) also plays a role in the generation of reactive oxygen species (ROS). Antioxidant defense systems encompass catalase, which catalyzes the conversion of hydrogen peroxide ($H_2O_2$) into water and oxygen, as well as vitamins C, E and β-carotene, which neutralize hydroxyl and singlet oxygen radicals, so averting oxidative damage. The equilibrium between reactive oxygen species (ROS) generation and detoxification in mitochondria is essential for cellular health; its disturbance may result in aging, degenerative illnesses and many pathologies.*

induce oxidative damage is primarily linked to their ability to interact with metal ions, light and oxygen. When carotenoids are exposed to these factors, they may undergo photooxidation or redox cycling [82]. Herein, carotenoids absorb photons, transferring their energy to molecular oxygen, which then produces singlet oxygen or other ROS. Further, at elevated concentrations, carotenoids fail to scavenge free radicals efficiently, leading to the generation of reactive intermediates capable of inducing oxidative stress. This phenomenon is exacerbated under high oxygen tension, where the excessive carotenoid presence can result in autoxidation [83].

The extended conjugated polyene chains of carotenoids make them susceptible to oxidation in the presence of light or transition metals [18]. β-Carotene and lutein can react with molecular oxygen or metal ions to produce singlet oxygen and lipid peroxides. These oxidative reactions, instead of neutralizing free radicals, amplify oxidative damage to cellular components. ROS can initiate lipid peroxidation, damage cellular membranes and disrupt mitochondrial function, leading to apoptosis or necrosis. Redox cycling, on the other hand, involves the repeated oxidation and reduction of carotenoids in the presence of metal ions like iron or copper, which can generate ROS, contributing to cellular damage (**Table 1**) [8, 84]. Such prooxidant behavior highlights the importance of optimal concentration and environmental conditions to harness their protective effects while minimizing potential oxidative harm.

The process by which lipid peroxyl radicals ($LOO^{\cdot}$) and ß-carotene (CAR) combine to generate a variety of carbon-centered radicals, including $(LOO)^{-2}$.

CAR-(OOL)) (LOO) -CAR-OOL·, LOO-CAR-OOL and LOO-CAR· is represented in equation:

$$CAR + LOO\bullet \underset{\beta\text{-carotene}}{-----------------------------} LOO - CAR\bullet \tag{1}$$

$$LOO - CAR\bullet + LOO \underset{\beta\text{-carotene}}{-----------} LOO - CAR - OOL \tag{2}$$

Carotenoids can disrupt the balance between antioxidants and prooxidants in the body. While carotenoids like β-carotene are known to induce antioxidant enzymes under normal conditions, they can act as prooxidants when cellular redox states are altered, or when antioxidants like vitamin E are insufficient. This shift can cause an increase in ROS production, particularly when carotenoid levels are high. In certain pathological states, such as cancer or cardiovascular diseases, this prooxidant activity may accelerate disease progression by promoting inflammation, DNA damage and tumorigenesis. Understanding the delicate balance between antioxidant and prooxidant functions of carotenoids is essential for maximizing their therapeutic potential while minimizing adverse effects.

## 3. Redox dynamics of carotenoids: Molecular mechanisms

The redox dynamics of carotenoids are central to their dual role as antioxidants and, under specific conditions, prooxidants. These compounds can directly scavenge ROS and quench singlet oxygen, effectively neutralizing oxidative stress and protecting cellular components. Their ability to interact with free radicals is largely dependent on their conjugated polyene structure, which allows the delocalization of unpaired electrons [82]. This antioxidant activity is influenced by factors such as oxygen tension, carotenoid concentration and the presence of other antioxidants. At physiological levels, carotenoids stabilize redox balance, reducing ROS accumulation and preventing lipid peroxidation [82]. However, under high concentrations or in environments with excessive ROS, carotenoids may adopt prooxidant behavior, facilitating oxidative chain reactions [85]. This shift occurs due to their interaction with ROS or lipid radicals, particularly when the antioxidant network is overwhelmed or unbalanced. Thus, the redox dynamics of carotenoids are tightly regulated, emphasizing the importance of maintaining their levels within physiological limits to support cellular homeostasis and mitigate oxidative damage.

### 3.1 Key pathways of redox cycling and ROS regulation

Carotenoids regulate redox dynamics primarily through their ability to quench ROS and modulate redox homeostasis. They efficiently neutralize singlet oxygen ($^1O_2$) and other reactive species by undergoing redox cycling, which involves electron transfer, hydrogen abstraction or physical quenching [85]. For instance, lycopene exhibits high singlet oxygen quenching efficiency, while β-carotene and lutein scavenge free radicals *via* their polyene chains. This redox activity prevents lipid peroxidation, protein oxidation and DNA damage in cellular systems [82]. Additionally, carotenoids can act as prooxidants under high oxygen tension, generating reactive intermediates that influence oxidative stress depending on cellular context.

### 3.2 Crosstalk with cellular signaling pathways and gene expression

As mentioned in Section 2, carotenoids interact with key cellular signaling pathways to regulate antioxidant responses and stress signaling. They activate nuclear factor erythroid 2–related factor 2 (Nrf2), a transcription factor that upregulates genes encoding antioxidant enzymes like glutathione peroxidase, SOD and catalase [71]. This cascade enhances cellular resilience against oxidative damage. Simultaneously, carotenoids suppress nuclear factor-kappa B (NF-κB), reducing inflammation and ROS-mediated damage. Lycopene, for example, has been shown in human studies to inhibit NF-κB activity, decreasing pro-inflammatory cytokine production in metabolic syndrome patients [86, 87]. These molecular interactions underscore the dual role of carotenoids as both direct antioxidants and modulators of gene expression.

### 3.3 Apoptosis, autophagy and carotenoid-mediated oxidative modulation

Carotenoids influence apoptosis and autophagic responses of the cell, the two critical processes for maintaining cellular homeostasis. By alleviating ROS levels, carotenoids like astaxanthin provide protection against oxidative stress-induced apoptosis in neuronal cells, as demonstrated in *in vitro* and *in vivo* studies [88]. Under specific situations, carotenoids can induce apoptosis *via* activating oxidative stress pathways, as well as mitochondrial or death receptor-mediated pathways. This way, carotenoids trigger apoptosis in cancer cells, hence facilitating tumor reduction [89, 90]. Moreover, carotenoids control autophagic flow, enhancing the clearance of damaged organelles and proteins. Zeaxanthin demonstrates the potential to subdue autophagic indicators in oxidative stress models, hence enhancing cytoprotection [91]. This dual modulation underscores their function as redox-sensitive regulators in both health and illness.

Carotenoids restore cellular homeostasis under stress by controlling autophagy *via* pathways including AMP-activated protein kinase (AMPK) and mechanistic target of rapamycin (mTOR) [92]. In clinical studies, dietary carotenoid supplementation has been linked to decreased markers of oxidative stress and apoptosis, especially in conditions such as cardiovascular disease and metabolic syndrome. Consequently, diets abundant in carotenoids have been associated with enhanced autophagic activity, particularly in older populations, potentially postponing age-related cellular deterioration [93]. This dual role in modulating apoptosis and autophagy underscores the importance of carotenoids as redox-sensitive regulators. Their context-dependent effects highlight their potential for therapeutic applications in managing oxidative stress-related diseases, including neurodegeneration, cancer and aging.

## 4. Translational applications of carotenoid bioactivity

Carotenoids hold significant promise in translational medicine owing to their antioxidant, anti-inflammatory and redox-regulatory properties, making them valuable for preventing and managing chronic diseases. Their ability to modulate apoptosis and autophagy positions them as potential adjuvants in cancer therapies, selectively targeting tumor cells while protecting healthy tissues.

### 4.1 Nutraceuticals and functional foods: Increasing dietary antioxidants

Because of their strong antioxidant qualities and related health advantages, carotenoids are being used more and more in functional foods and nutraceuticals. ROS can be scavenged and oxidative damage can be decreased by these bioactive chemicals, which are found in large quantities in fruits, vegetables and marine sources. Carotenoids such as lutein, β-carotene and lycopene are used in functional food formulations to enhance skin, eye and cardiovascular health. According to clinical research, diets high in carotenoid-rich foods increase blood antioxidant capacity and lower oxidative stress indicators. Drinks fortified with lutein have been demonstrated to increase the optical density of macular pigments, which may reduce the risk of AMD. These developments emphasize the importance of carotenoids in contemporary dietary therapies.

### 4.2 Therapeutic potential in managing oxidative stress-associated disorders

Carotenoids are regarded as adjunctive therapy owing to their therapeutic efficacy in addressing oxidative stress-related conditions, including cancer, neurological disorders and cardiovascular illnesses. Research has shown that lycopene decreases MDA, improves lipid profiles and mitigates oxidative damage in persons with metabolic syndrome. Astaxanthin may traverse the blood-brain barrier and mitigate ROS-induced neuronal damage, demonstrating neuroprotective advantages in conditions such as Alzheimer's disease. Research indicates that β-carotene supplementation reduced oxidative markers and inflammation in individuals with rheumatoid arthritis, suggesting that carotenoids also affect inflammatory pathways. These therapeutic advantages endorse its potential application in adjunctive therapy methods.

Despite the fact that carotenoid supplementation has potential health advantages, its safety and efficacy warrant a careful evaluation in clinical contexts (**Figure 4**). Clinical research underscores the need for proper doses to mitigate prooxidant dangers and enhance therapeutic advantages. Furthermore, formulations that integrate

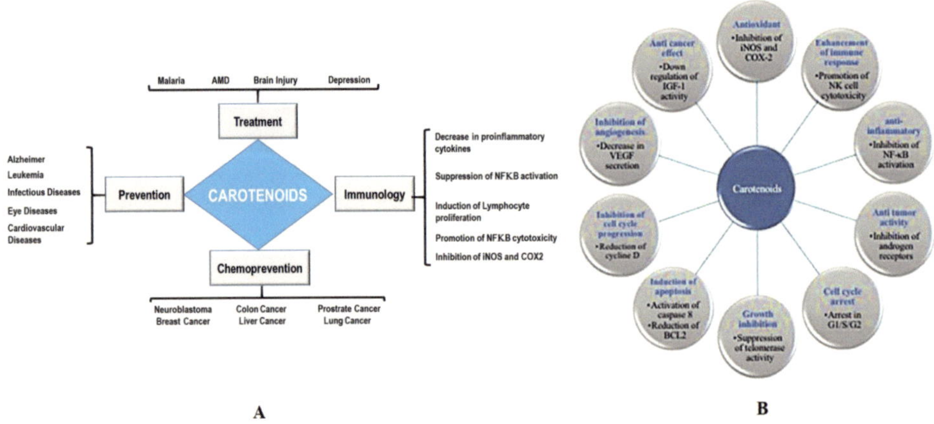


**Figure 4.**
*(A) Multifaceted roles of carotenoids in disease prevention, treatment and immunomodulation, including their effects on cancer, neurodegenerative and infectious diseases. (B) Mechanistic actions of carotenoids, such as antioxidant activity, anti-inflammatory effects, immune regulation and cell cycle modulation.*

carotenoids with synergistic nutrients, like vitamins C and E, have shown improved absorption and effectiveness. Research indicates that carotenoid-enriched soft gels normalize oxidative stress indicators in non-alcoholic fatty liver disease (NAFLD) without eliciting deleterious consequences. These findings highlight the significance of evidence-based methodologies in clinical practice.

### 4.3 Bridging research and clinical utility

The incorporation of carotenoids into healthcare frameworks requires a multidisciplinary strategy, merging knowledge from molecular biology, pharmacology and clinical nutrition. Innovative technologies, including nanocarriers and bioengineering, are utilized to improve carotenoid bioavailability and stability, therefore expanding their therapeutic relevance. Furthermore, progress in omics technology enhances comprehension of individual reactions to carotenoid supplementation, enabling the development of individualized nutrition plans. Longitudinal investigations and extensive clinical trials are essential for determining the long-term safety, effectiveness and appropriate use of carotenoids in disease prevention and therapy (**Figure 4**). These advancements indicate a potential for the extensive use of carotenoid-based therapies in clinical and public health sectors.

## 5. Innovations and future directions in carotenoid research

Recent advances in carotenoid metabolism have brought attention to the intricate interactions that these compounds have with cellular signaling and redox mechanisms [4]. Targeted therapies that take advantage of these interactions to prevent chronic diseases linked to oxidative stress have been made possible by advancements in the use of mass spectrometry and isotopic tracers, nanotechnology and bioinformatics, which have also allowed for a deeper understanding of the *in vivo* conversion of carotenoids and their impact on cellular redox states [94].

### 5.1 Cutting-edge techniques for investigating carotenoid bioactivity

Investigation of carotenoid bioactivity has advanced due to sophisticated analytical methods that provide improved accuracy in elucidating their physiological functions. High-resolution mass spectrometry (HRMS) and liquid chromatography-tandem mass spectrometry (LC-MS/MS) have facilitated the precise identification and quantification of carotenoid metabolites in biological tissues. Moreover, the application of sophisticated imaging methodologies, including confocal microscopy, has enabled researchers to monitor carotenoid dispersion and localization inside certain cellular compartments in real time. These improvements have yielded novel insights into the interactions of carotenoids with lipophilic membranes, their impact on gene expression and their modulation of cellular processes. Methods like CRISPR-Cas9 gene editing and RNA sequencing have been utilized to investigate the molecular processes by which carotenoids affect cellular signaling pathways and metabolic networks, enhancing our comprehension of their therapeutic potential.

### 5.2 Future prospects: Personalized approaches in nutrition and therapeutics

Carotenoid research offers the potential for personalized nutrition and therapeutics as individual differences in carotenoid metabolism are becoming more recognized. Genetic polymorphisms in carotenoid-metabolizing enzymes can modulate the bioavailability and effectiveness of carotenoids [7]. Nutrigenomics is advancing to determine ideal carotenoid consumption levels based on an individual's genetic composition and health state. Personalized carotenoid supplementation can be used in clinical practice to target specific oxidative stress-related conditions like cancer, AMD and cardiovascular disease [95].

## 6. Conclusion

In summary, carotenoids provide a double-edged sword in health management; hence, their advantages must be harnessed judiciously. By bridging the gap between basic research and real-world application, scientists and medical professionals can fully unlock the potential of carotenoids to prevent illnesses linked to oxidative stress and improve human health. Nonetheless, the realization of their full potential as therapeutic and functional agents depends heavily on strategic, evidence-based approaches. Future research should focus on uncovering the molecular mechanisms behind carotenoid redox dynamics, their effects on cellular signaling and gene expression and their therapeutic applications. Advances in biotechnology, nanotechnology and analytical tools will be pivotal in unlocking their full potential in health and disease management.

## Author details

Aisha Farhana[1*], Yusuf Saleem Khan[2], Abdullah Alsrhani[1], Emad Manni[1], Ayman A.M. Alameen[1], Wassila Derafa[3], Nada Alhathlaul[3], Muhammad Atif[1] and Lienda Bashier Eltayeb[4]

1 Department of Clinical Laboratory Sciences, College of Applied Medical Sciences, Jouf University, Saudi Arabia

2 Department of Anatomy, College of Medicine, University of Hail, Saudi Arabia

3 Department of Chemistry, College of Science, Jouf University, Saudi Arabia

4 Department of Medical Laboratory Sciences, College of Applied Medical Sciences, Prince Sattam Bin Abdulaziz University, Riyadh, Saudi Arabia

*Address all correspondence to: afarhana@ju.edu.sa

IntechOpen

© 2025 The Author(s). Licensee IntechOpen. This chapter is distributed under the terms of the Creative Commons Attribution License (http://creativecommons.org/licenses/by/4.0), which permits unrestricted use, distribution, and reproduction in any medium, provided the original work is properly cited.

## References

[1] Yabuzaki J. Carotenoids database: Structures, chemical fingerprints and distribution among organisms. Database. 2017;**2017**:bax004

[2] Deis L, Quiroga AM, De Rosas MI. Coloured compounds in fruits and vegetables and health. In: Psychiatry and Neuroscience Update: From Epistemology to Clinical Psychiatry–Vol. IV. Cham, Switzerland: Springer Nature; 2021. pp. 343-358

[3] Olmedilla-Alonso B. Carotenoid markers of dietary exposure and nutritional status. Nutrients. 2023;**15**:2359

[4] Martini D et al. What is the current direction of the research on carotenoids and human health? An overview of registered clinical trials. Nutrients. 2022;**14**(6):1191

[5] Crupi P et al. Overview of the potential beneficial effects of carotenoids on consumer health and well-being. Antioxidants. 2023;**12**(5):1069

[6] Lu W et al. Antioxidant activity and healthy benefits of natural pigments in fruits: A review. International Journal of Molecular Sciences. 2021;**22**(9):4945

[7] Bas TG. Bioactivity and bioavailability of carotenoids applied in human health: Technological advances and innovation. International Journal of Molecular Sciences. 2024;**25**(14):7603

[8] Lucas M et al. Antioxidant and pro-oxidant activities of carotenoids. In: Plant Antioxidants and Health. Cham: Springer; 2020. pp. 1-27

[9] Gao Q et al. Antioxidant and prooxidant activities of phenolic acids commonly existed in vegetables and their relationship with structures. Food Science and Technology. 2022;**42**:e07622

[10] Timoshnikov VA et al. Mechanistic insights of chelator complexes with essential transition metals: Antioxidant/pro-oxidant activity and applications in medicine. International Journal of Molecular Sciences. 2022;**23**(3):1247

[11] Rajashekar CB. Dual role of plant phenolic compounds as antioxidants and prooxidants. American Journal of Plant Sciences. 2023;**14**(1):15-28

[12] Esteban R et al. Versatility of carotenoids: An integrated view on diversity, evolution, functional roles and environmental interactions. Environmental and Experimental Botany. 2015;**119**:63-75

[13] Harrison EH, Curley RW. Carotenoids and retinoids: Nomenclature, chemistry, and analysis. Sub-Cellular Biochemistry. 2016;**81**:1-19

[14] Kabir MT et al. Therapeutic promise of carotenoids as antioxidants and anti-inflammatory agents in neurodegenerative disorders. Biomedicine & Pharmacotherapy. 2022;**146**:112610

[15] Husain N, Kumar A. Reactive oxygen species and natural antioxidants: A review. Advances in Bioresearch. 2012;**3**(4):164-175

[16] Prakash D, Gupta C. 12 Carotenoids: Chemistry and health benefits. In: Phytochemicals of Nutraceutical Importance. Wallingford, UK: CABI; 2014. p. 181

[17] Riaz M, Zia-Ul-Haq M, Dou D. Chemistry of carotenoids. In:

Carotenoids: Structure and Function in the Human Body. Cham, Switzerland: Springer Nature; 2021. pp. 43-76

[18] Polyakov NE et al. The endless world of carotenoids—Structural, chemical and biological aspects of some rare carotenoids. International Journal of Molecular Sciences. 2023;**24**(12):9885

[19] Llansola-Portoles MJ, Pascal AA, Robert B. Electronic and vibrational properties of carotenoids: From in vitro to *in vivo*. Journal of the Royal Society Interface. 2017;**14**(135):20170504

[20] Schlücker S et al. Density functional and vibrational spectroscopic analysis of β-carotene. Journal of Raman Spectroscopy. 2003;**34**(6):413-419

[21] Ashikhmin AA et al. ζ-carotene: Generation and quenching of singlet oxygen, comparison with phytofluene. Biochemistry (Moscow). 2022;**87**(10):1169-1178

[22] Meléndez-Martínez AJ et al. Relationship between the colour and the chemical structure of carotenoid pigments. Food Chemistry. 2007;**101**(3):1145-1150

[23] Kwatra B. A review on potential properties and therapeutic applications of lycopene. International Journal of Medical and Biomedical Studies. 2020;**4**(10):32553

[24] Thomas SE, Johnson EJ. Xanthophylls. Advances in Nutrition. 2018;**9**(2):160-162

[25] Zia-Ul-Haq M. Historical and introductory aspects of carotenoids. In: Carotenoids: Structure and Function in the Human Body. Cham, Switzerland: Springer Nature; 2021. pp. 1-42

[26] Demmig-Adams B, Polutchko SK, Adams WW III. Structure-function-environment relationship of the isomers zeaxanthin and lutein. Photochemistry. 2022;**2**(2):308-325

[27] Bouyahya A et al. Sources, health benefits, and biological properties of zeaxanthin. Trends in Food Science & Technology. 2021;**118**:519-538

[28] Nair A et al. Astaxanthin as a king of ketocarotenoids: Structure, synthesis, accumulation, bioavailability and antioxidant properties. Marine Drugs. 2023;**21**(3):176

[29] Brotosudarmo THP et al. Structures of astaxanthin and their consequences for therapeutic application. International Journal of Food Science. 2020;**2020**(1):2156582

[30] Liu X et al. Recent advances in health benefits and bioavailability of dietary astaxanthin and its isomers. Food Chemistry. 2023;**404**:134605

[31] Bungau S et al. Health benefits of polyphenols and carotenoids in age-related eye diseases. Oxidative Medicine and Cellular Longevity. 2019;**2019**(1):9783429

[32] Yamagata K. Carotenoids regulate endothelial functions and reduce the risk of cardiovascular disease. In: Carotenoids. Vol. 25(106.10). London, UK: IntechOpen; 2017. p. 5772

[33] Miller AP. Discovering carotenoid biodistribution pathways: novel Mechanisms and their implications for health [thesis]. USA: University of Illinois at Urbana-Champaign; 2023

[34] Bowen PE, Stacewicz-Sapuntzakis M, Diwadkar-Navsariwala V. Carotenoids in human nutrition. In: Pigments in Fruits and Vegetables: Genomics and Dietetics. New York, NY: Springer; 2015. pp. 31-67

[35] Kamal S et al. Eye sight and carotenoids. In: Carotenoids: Structure and Function in the Human Body. Cham, Switzerland: Springer Nature; 2021. pp. 609-647

[36] Thirunavukarasu AJ, Ross AC, Gilbert RM. Vitamin A, systemic T-cells, and the eye: Focus on degenerative retinal disease. Frontiers in Nutrition. 2022;**9**:914457

[37] Davinelli S et al. Astaxanthin as a modulator of Nrf2, NF-κB, and their crosstalk: Molecular mechanisms and possible clinical applications. Molecules. 2022;**27**(2):502

[38] Hadad N, Levy R. The synergistic anti-inflammatory effects of lycopene, lutein, β-carotene, and carnosic acid combinations via redox-based inhibition of NF-κB signaling. Free Radical Biology and Medicine. 2012;**53**(7):1381-1391

[39] Pandey P et al. Targeting Wnt/β-catenin pathway by flavonoids: Implication for cancer therapeutics. Nutrients. 2023;**15**(9):2088

[40] Sevastre A-S et al. Intracellular pathways and mechanisms of colored secondary metabolites in cancer therapy. International Journal of Molecular Sciences. 2022;**23**(17):9943

[41] Kaulmann A, Bohn T. Carotenoids, inflammation, and oxidative stress—Implications of cellular signaling pathways and relation to chronic disease prevention. Nutrition Research. 2014;**34**(11):907-929

[42] Bohn T et al. Mechanistic aspects of carotenoid health benefits–where are we now? Nutrition Research Reviews. 2021;**34**(2):276-302

[43] Brtko J, Dvorak Z. Natural and synthetic retinoid X receptor ligands and their role in selected nuclear receptor action. Biochimie. 2020;**179**:157-168

[44] Purohith R, Nagalingaswamy NP, Shivananju NS. Dietary carotenoids in managing metabolic syndrome and role of PPARs in the process. Current Nutrition & Food Science. 2020;**16**(6):846-853

[45] Yang G et al. Astaxanthin suppresses oxidative stress and calcification in vertebral cartilage endplate via activating Nrf-2/HO-1 signaling pathway. International Immunopharmacology. 2023;**119**:110159

[46] Wu T et al. β-carotene protects mice against lipopolysaccharide and D-galactosamine induced acute liver injury via regulation of NF-κB, MAPK, and Nrf2 signaling. Journal of Oleo Science. 2023;**72**(11):1027-1035

[47] Dovinova I et al. The role of Nrf2 and PPARγ in the improvement of oxidative stress in hypertension and cardiovascular diseases. Physiological Research. 2020;**69**(Suppl 4):S541

[48] Bohn T. Carotenoids and markers of oxidative stress in human observational studies and intervention trials: Implications for chronic diseases. Antioxidants. 2019;**8**(6):179

[49] Savory JG et al. Identification of novel retinoic acid target genes. Developmental Biology. 2014;**395**(2):199-208

[50] Kim J-S et al. Red paprika (*Capsicum annuum* L.) and its main carotenoids, capsanthin and β-carotene, prevent hydrogen peroxide-induced inhibition of gap-junction intercellular communication. Chemico-Biological Interactions. 2016;**254**:146-155

[51] Bertram JS. Cancer chemoprevention by retinoids and carotenoids: Proposed

role of gap junctional communication. In: Vitamins and Minerals in the Prevention and Treatment of Cancer. Boca Raton, FL: CRC Press; 2018. pp. 31-50

[52] Hammerling U, Kim Y-K, Quadro L. Quantum chemistry rules retinoid biology. Communications Biology. 2023;**6**(1):227

[53] Sabuncuoglu S, Yalcinkaya A, Blumenberg M. Importance of Oxidative Stress and Antioxidant System in Health and Disease. BoD–Books on Demand; 2023

[54] Kumar S et al. Astaxanthin: A super antioxidant from microalgae and its therapeutic potential. Journal of Basic Microbiology. 2022;**62**(9):1064-1082

[55] Deng J-J et al. Enzymatic conversion and recovery of protein, chitin, and astaxanthin from shrimp shell waste. Journal of Cleaner Production. (Amsterdam: Elsevier). 2020;**271**:122655

[56] Davan I et al. Astaxanthin as a potent antioxidant for promoting bone health: An up-to-date review. Antioxidants. 2023;**12**(7):1480

[57] Gao F et al. Astaxanthin provides neuroprotection in an experimental model of traumatic brain injury via the Nrf2/HO-1 pathway. American Journal of Translational Research. 2021;**13**(3):1483

[58] Li X et al. Protective effects of astaxanthin supplementation against ultraviolet-induced photoaging in hairless mice. Biomedicine. 2020;**8**(2):18

[59] Sekikawa T et al. Effects of diet containing astaxanthin on visual function in healthy individuals: A randomized, double-blind, placebo-controlled, parallel study. Journal of Clinical Biochemistry and Nutrition. 2022;**72**(1):74

[60] Zhou X et al. Systematic review and meta-analysis on the effects of astaxanthin on human skin ageing. Nutrients. 2021;**13**(9):2917

[61] Gammone MA, Danese A, D'Orazio N. Anti-angiogenetic agents from the sea: A new potential preventive and therapeutic wave? Anti-Cancer Agents in Medicinal Chemistry-Anti-Cancer Agents. 2020;**20**(17):2005-2011

[62] Mohiti S et al. Spirulina supplementation and oxidative stress and pro-inflammatory biomarkers: A systematic review and meta-analysis of controlled clinical trials. Clinical and Experimental Pharmacology and Physiology. 2021;**48**(8):1059-1069

[63] Hozawa A et al. Study profile of the Tohoku medical megabank community-based cohort study. Journal of Epidemiology. 2021;**31**(1):65-76

[64] Tanaka T, Shnimizu M, Moriwaki H. Cancer chemoprevention by caroteno. Molecules. 2012;**17**(3):3202-3242

[65] Marini A, Krutmann J. Beauty from the inside: Nutrition and skin ageing. In: Anti-Ageing Nutrients: Evidence-Based Prevention of Age-Associated Diseases. Hoboken, NJ: John Wiley & Sons; 2015. pp. 299-314

[66] Takashima M et al. Capacity of peroxyl radical scavenging and inhibition of lipid peroxidation by β-carotene, lycopene, and commercial tomato juice. Food & Function. 2012;**3**(11):1153-1160

[67] Kazaleh M et al. Oxidative stress and the pathogenesis of aortic aneurysms. Biomedicine. 2023;**12**(1):3

[68] Al-Tarifi BY et al. Application of astaxanthin and its lipid stability in bakery product. Current Research in

Nutrition and Food Science Journal. 2020;**8**(3):962-974

[69] Ahn YJ, Kim H. Lutein as a modulator of oxidative stress-mediated inflammatory diseases. Antioxidants. 2021;**10**(9):1448

[70] Turkler C et al. The ameliorative effect of lutein on ovarian ischemia-reperfusion injury in rats. Bratislavske lekarske listy. 2018;**119**(11):713-717

[71] Li H et al. Lutein suppresses oxidative stress and inflammation by Nrf2 activation in an osteoporosis rat model. Medical Science Monitor: International Medical Journal of Experimental and Clinical Research. 2018;**24**:5071

[72] Lindbergh CA et al. Relationship of lutein and zeaxanthin levels to neurocognitive functioning: An fMRI study of older adults. Journal of the International Neuropsychological Society. 2017;**23**(1):11-22

[73] Göncü T et al. Anti-inflammatory effect of lycopene on endotoxin-induced uveitis in rats. Arquivos Brasileiros de Oftalmologia. 2016;**79**:357-362

[74] Tamilselvan P et al. Lycopene, a carotenoid antioxidant against bisphenol a (BPA) motivated experimental male infertility. Journal of Biomedical and Therapeutic Sciences. 2014;**1**(1):41-47

[75] Li M et al. Lycopene as a therapeutic agent against aflatoxin B1-related toxicity: Mechanistic insights and future directions. Antioxidants. 2024;**13**(4):452

[76] Murillo AG, Hu S, Fernandez ML. Zeaxanthin: Metabolism, properties, and antioxidant protection of eyes, heart, liver, and skin. Antioxidants. 2019;**8**(9):390

[77] El-Akabawy G, El-Sherif NM. Zeaxanthin exerts protective effects on acetic acid-induced colitis in rats via modulation of pro-inflammatory cytokines and oxidative stress. Biomedicine & Pharmacotherapy. 2019;**111**:841-851

[78] Jiménez-Escrig A et al. Evaluation of free radical scavenging of dietary carotenoids by the stable radical 2, 2-diphenyl-1-picrylhydrazyl. Journal of the Science of Food and Agriculture. 2000;**80**(11):1686-1690

[79] Krinsky NI. Singlet oxygen in biological systems. Trends in Biochemical Sciences. 1977;**2**(2):35-38

[80] Britton G. Getting to know carotenoids. In: Methods in Enzymology. Amsterdam: Elsevier; 2022. pp. 1-56

[81] Metibemu DS, Ogungbe IV. Carotenoids in drug discovery and medicine: Pathways and molecular targets implicated in human diseases. Molecules. 2022;**27**(18):6005

[82] Horn PJ. Where do the electrons go? How numerous redox processes drive phytochemical diversity: Redox processes in phytochemistry. Phytochemistry Reviews. 2021;**20**(2):367-407

[83] Black HS et al. The benefits and risks of certain dietary carotenoids that exhibit both anti-and pro-oxidative mechanisms—A comprehensive review. Antioxidants. 2020;**9**(3):264

[84] Boehm F, Edge R, Truscott TG. Photochemical and photophysical properties of carotenoids and reactive oxygen species: Contradictions relating to skin and vision. Oxygen. 2023;**3**(3):322-335

[85] Kuczyńska M, Jakubek P, Bartoszek A. More than just antioxidants: Redox-active components and mechanisms shaping redox

signalling network. Antioxidants. 2022;**11**(12):2403

[86] Simone RE et al. Lycopene inhibits NF-kB-mediated IL-8 expression and changes redox and PPARγ signalling in cigarette smoke–stimulated macrophages. PLoS One. 2011;**6**(5):e19652

[87] Wu S et al. Study of the anti-inflammatory mechanism of β-carotene based on network pharmacology. Molecules. 2023;**28**(22):7540

[88] Park H-A et al. Anti-apoptotic effects of carotenoids in neurodegeneration. Molecules. 2020;**25**(15):3453

[89] Saini RK et al. Dietary carotenoids in cancer chemoprevention and chemotherapy: A review of emerging evidence. Pharmacological Research. 2020;**157**:104830

[90] Baeza-Morales A et al. The antitumour mechanisms of carotenoids: A comprehensive review. Antioxidants. 2024;**13**(9):1060

[91] Pradhan B et al. Zeaxanthin: A novel therapeutic agent against human diseases. In: Algal Biotechnology. Boca Raton, FL: CRC Press; 2024. pp. 307-319

[92] Patra S et al. Autophagy-modulating phytochemicals in cancer therapeutics: Current evidences and future perspectives. In: Seminars in Cancer Biology. Amsterdam: Elsevier; 2022

[93] Bakac ER et al. A narrative review: The effect and importance of carotenoids on aging and aging-related diseases. International Journal of Molecular Sciences. 2023;**24**(20):15199

[94] Britton G. Carotenoid research: History and new perspectives for chemistry in biological systems. Biochimica et Biophysica Acta (BBA)-Molecular and Cell Biology of Lipids. 2020;**1865**(11):158699

[95] Tragni V et al. Personalized medicine in mitochondrial health and disease: Molecular basis of therapeutic approaches based on nutritional supplements and their analogs. Molecules. 2022;**27**(11):3494

Chapter 4

# Antioxidant Properties of Natural Bioactive Compounds

*Okram Abemsana Devi and Mridula Saikia Barooah*

## Abstract

Bioactive compounds are extra-nutritional or non-essential ingredients that are normally found in trace amounts in food and appear to have positive health benefits. Secondary metabolites, a term used to refer to the bioactive components generated from plants, are recognized to have promising medicinal qualities, particularly antioxidative ones. Plants contain a variety of bioactive molecules, including flavonoids, polyphenols, tannins, coumarins, terpenoids, alkaloids, saponins, anthocyanins, carotenoids, and plant sterols. These substances have the ability to lessen the negative effects of harmful free radicals, which may contribute to the premature death of immune cells and many more. This chapter examines the role of diverse bioactive compounds and their derivatives that act as antioxidants in preventing various diseases. Despite the positive track record of natural products, more research is needed to fully comprehend the potential role that natural bioactive compounds could play as a significant source of novel therapeutic agents.

**Keywords:** bioactive compounds, secondary metabolites, antioxidants, free radicals, diseases, antioxidant activity

## 1. Introduction

### 1.1 An overview of bioactive compounds

Each and every living thing, from single cell bacteria to a million cell plants, processes a wide range of chemical substances in order to survive and subsistence. There are two general categories into which all chemicals in a biological system can be placed. The chemicals targeted at growth and development are called primary metabolites, and they comprise purines, pyrimidines, amino acids, carbohydrates, and proteins with nucleic acids. Secondary metabolites are a class of compounds other than primary metabolites that are thought to help plants interact with their environment and increase their overall capacity to survive and overcome local challenges. These compounds include lignans, alkaloids, terpenes, flavonoids, phenolics, curcumin, saponin, glucosides, and steroids [1]. Certain chemicals are discovered to serve as defense mechanisms against pathogens, predators, UV rays, parasites, and oxidants. Additionally, they are shown to aid in reproductive activities by acting as appealing fragrances and coloring agents, as well as to promote competition

 IntechOpen

among different species. Plant-derived bioactive compounds are known as secondary metabolites that have desired health and wellness benefits in humans and animals [2]. A number of investigations [3, 4] have suggested that 5000–25,000 distinct phytochemicals present in natural foods may have bioactive qualities. The health benefits of consuming certain foods have been widely known since ancient times. Nonetheless, during the past few decades, research on foods as a source of beneficial bioactive chemicals has garnered attention for a variety of purposes [5, 6]. The term "bioactive compounds" is still imprecise and vague despite a great deal of important research in many domains.

Food bioactive compounds are secondary metabolites that are non-essential or extra-nutritional constituents which typically occur in small quantities in foods, having beneficial health effects [7]. It also refers to all compounds that are mostly without nutritional value and naturally present in food that exert a certain bioactive effect on the human body [8–10]. As the name suggests, the bioactive substances and components are defined by a certain biological activity that directly affects living things (Greek "bios" means life, and Latin "actives" means dynamic or full of enegy) [11]. Plant-derived bioactive components originating from metabolism are generally known as secondary metabolites and have promising therapeutic attributes, especially antioxidative properties [12]. Though usually in minute amounts, they can be found in a wide range of plant sources including vegetables, fruits, grains, and legumes [12]. Phenolic polyphenols, flavonoids, flavones, flavonols, tannins, quinones, coumarins, terpenoids, alkaloids, essential oils, lectins, carotenoids, anthocyanins, betalains, glucosinolates, and plant sterols are some of the significant families of active substances present in plants.

There is a severe paucity of knowledge on the precise biological process by which plant-based bioactive dietary components provide health-promoting advantages. Food ingredients that are bioactive work at several or related target sites at the same. Bioactive food elements have been demonstrated to lower the risk of infection, osteoporosis, cardiovascular disease, and cancer [13]. Bioactive ingredients offer health-promoting properties in various stages of illness development that are connected with several development stages, from onset to advancement. For example, through binding cholesterol in the digestive tract, isoflavones can reduce circulating oxidized low-density plasma lipoproteins in cardiovascular disorders [14]. Additionally, it elevates bile excretion, lowers endogenous cholesterol levels, and modulates arterial elasticity, which enhances blood vessel dilation and constriction responses [15, 16]. It also reduces the absorption of cholesterol from diet [7, 17]. Spices and herbs have the ability to prevent inflammation, which is a major contributing factor to a number of acute or chronic pathological conditions that lead to carcinogenesis or cardiovascular diseases. This is due to their antioxidant and anti-inflammatory properties [17]. Polyphenols, tocopherols, carotenoids, and allyl sulphides all work as antioxidants against harmful free radicals and reactive oxygen species (ROS). Antioxidants primarily function by regulating the redox potential and depleting possible initiators of carcinogenesis. Different chemical reactions, such as the induction of apoptosis (cell death) and the activation of signal transduction, can be brought on by redox changes within a cell [14]. Thus, redox and antioxidant regulation of physiological and pathological processes is crucial for maintaining health and preventing disease. Additionally, it has been demonstrated that bioactive food components promote the growth of certain immune cell types and thephagocytic activity of macrophages, hence boosting disease resistance [14].

## 2. Major bioactive compounds

Plants contain a diverse range of biologically active secondary metabolites that have been isolated, identified, and used in food products such as food additives, flavors, nutraceuticals, coloring agents, and cosmetics. The pharmaceutical industry has also exclusively used these metabolites to discover and develop new medications. Numerous chemicals are produced by plants, including aquatic macrophytes, such as phenolic compounds, polyketides, alkaloids, peptides, flavonoids, steroids, terpenes, tannins, quinones, and coumarins (**Figure 1**). These compounds are used in medicine for their antioxidant, inflammatory, anticancer, antibacterial, antiviral, cytotoxic, and many other properties. They are also widely used in other disciplines, including scientific research, agriculture, human therapy, veterinary care, and medicine.

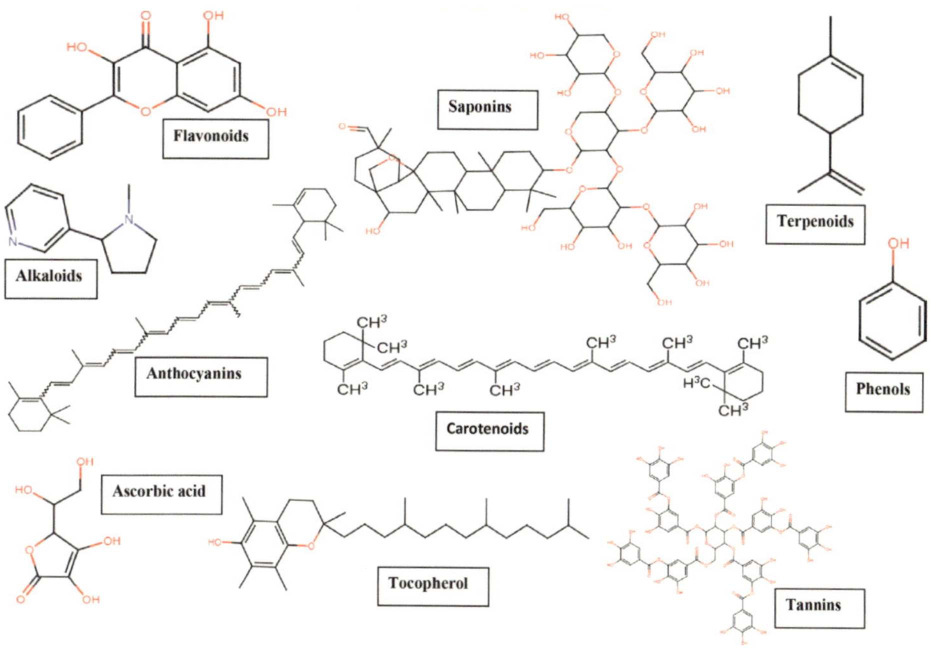


**Figure 1.**
*Structures of some common bioactive compounds.*

## 3. Phenols

Due to the presence of a hydroxyl (OH) group that is directly attached to an aromatic ring, phenols are a subclass of aromatic bioactive compounds that are weakly acidic. They are soluble in acetone, alcohol, and water and form precipitates that contain proteins [18]. According to Uttara et al. [19] and Ramaswamy [20], polyphenols offer protection against a variety of diseases that are linked to oxidative stress and damage caused by free radicals, including cancer, diabetes, autoimmune disorders, inflammatory diseases, and cardiovascular and neurodegenerative illness. The number and relative positions of hydroxyl groups in the aromatic ring structure, as well as the acid moiety, are factors that influence the antioxidant activity of phenolic compounds.

As hydroxycinnamic acids in phenolic compounds stimulate the delocalization of phenoxyl radicals, they are more potent antioxidants than hydroxybenzoic acids [21]. The highest antioxidant activity is conferred by hydroxylation of phenolic compounds in the 2- and 4-positions or in the 3-, 4-, and 5-positions. Consequently, the capacity to donate electrons will boost antioxidant activity. Due to increased electron delocalization, hydroxycinnamic acid esters, such as chorogenic acid, p-coumaroyltartaric acid, and caffeoyltaric acid, display higher levels of antioxidant activity than their parent hydroxycinnamic acids [22]. In addition to providing protection against oxidative damage, polyphenols are effective in lowering the risk of cancer, diabetes, hypertension, and coronary heart disease [23–25]. According to research, consuming high concentration of fruits and vegetable juice at least three times a week, can help prevent oxidative damage and delay the onset of Alzheimer's disease [26]. This lowers the risk of diabetes, hypertension, coronary heart disease, and cancer [25, 27]. According to numerous researches studies, the consumption of polyphenols reduces the prevalence of coronary heart diseases [28–30]. Polyphenols are thought to play a major role in the pathogenesis of atherosclerosis due to strong inhibitors of LDL oxidation [31]. Polyphenols may also have anti-inflammatory, antioxidant, and anti-platelet properties in addition to raising HDL levels as protective effects against cardiovascular illnesses [32]. Additionally, polyphenols might help to keep the atheroma plaque in stable form.

## 4. Flavonoids

Plants produce flavonoids, which are known as hydroxylated phenolic compounds, in reaction to microbial diseases. Flavonoids have a significant role in growth, cell differentiation, and defense signaling in plants, and they also aid in the responses of plants to severe abiotic challenges [33, 34]. The chemical nature of flavonoids depends on their degree of hydroxylation, degree of polymerization, structural class, other substitutions, and conjugations [35]. Based on the sites and numbers of hydroxyl groups, flavonoids are divided into various sub-classes such as flavonols (e.g., quercetin, myricetin, kaempferol, and fisetin), flavanones (e.g., naringenin, hesperetin, and flavanone), and flavones (e.g., apigenin, flavone, and luteolin) [33, 36]. Regardless of similar chemical properties, flavonoids are also involved in photosynthesis, growth, and protective biological systems including pollination, seed development, light screening, temperature acclimation, protection against UV damage, defense against pathogens, and drought resistance. Additionally, some flavonoids can act as signaling molecules for allelopathic interactions, including detoxifying agents and phytoalexins [37, 38]. Apart from the inherent bioactive characteristics of plants, flavonoids are phytochemicals that possess numerous biological activities that are therapeutically significant. These activities include antifungal, antiviral, antibacterial, antiosteoporotic, antiulcer, antihepatotoxic, and antiproliferative capabilities [39].

Typically, the flavonoids in food are responsible for its color, flavor, and ability to keep fat from oxidizing as well as preserving vitamins and enzymes [40]. When it comes to the uses of flavonoids in human nutrition, their ability to act as antioxidants is crucial in inhibiting the production of reactive oxygen and nitrogen species, which damage proteins, lipids, DNA, and other macromolecules [41, 42]. Due to their strong antioxidant potential both in vivo and in vitro, flavonoids as a dietary component offer health-promoting qualities [43, 44]. The antioxidant capacity of flavonoids is influenced by the way their functional groups are arranged in the nuclear structure. Many mechanisms of antioxidant activity, including radical scavenging and metal ion chelation capacity, are significantly influenced by the configuration, substitution, and total quantity of hydroxyl groups [35, 45]. Flavonoids' structures, especially the B ring with 3′, 4′ dihydroxy groups and the 3-OH group at the C ring, are key features for strong antioxidant activity as they can act as proton donors. The presence of a B-ring catechol group (dihydroxylated B-ring) is the primary marker of a flavonoid molecule's capacity to donate hydrogen. According to Schroeter et al. [46], flavonoids prevent lipids and other molecules from being oxidized by donating hydrogen atoms to ROO radicals. Flavonoids such as quercetin, catechin, and myricetin are strong free radical scavengers due to the presence of double bonds in hydroxyl positions in their molecules, which determine their antioxidant properties. Because of their unique chemical structure, they can reduce oxidative stress through a variety of mechanisms. According to Pietta [47], one of the mechanisms involved is that flavonoids scavenge superoxide, peroxyl, alkoxyl, and hydroxyl radicals in addition to reducing low-density lipoprotein oxidation. They also operate as chain-breaking antioxidants. On the other hand, flavonoids also function as enzyme inhibitors (inhibiting the enzymes that generate ROS: xanthine oxidase, protein kinase C, cyclooxygenase, lipoxygenase, glutathione S-transferase, microsomal monooxygenase, mitochondrial succinoxidase, and NADH oxidase) and metal chelators (lowering ferric and cupric ions). Flavonoids inhibit platelet aggregation, lowering the risk of cancer, heart disease, and thrombosis [48]. Sharma and Gujral [49] also stated that flavonoids play an important role in preventing a number of ROS-mediated disorders such as diabetes, high blood pressure, allergies, tumors, inflammation, ulcers, and heart diseases.

## 5. Terpenoids

Terpenoids, sometimes referred to as isoprenoids or terpenes, are one of the most prevalent and extensive families of natural products, with over 80,000 distinct chemicals involved in both primary and secondary metabolisms [50]. The term "terpene" was given to the compounds isolated from terpentine, a volatile liquid isolated from pine trees. Most terpenoids are water-insoluble, hydrophobic, and easily soluble in nonpolar organic solvents. They are readily hydrogenated, oxidized, polymerized, isomerized, and halogenated [51]. Terpenoids are ubiquitous in nature and have a range of biological activities, such as antioxidant, anti-inflammatory, anticancer, antifungal, antiviral, antiarrhythmic, antibacterial, antispasmodic, and antihistaminic activities [52]. Terpenoids are known to exhibit biological activity due to their lipophilicity and ability to partition into cellular membranes. They have the ability to interact with membrane-attached proteins within the cells or compromise the integrity of the membrane [53]. Likewise, Thenmozhi and Ramalakshmi [54] also stated that the presence of terpenoids may disrupt the cell membrane of bacteria, form a complex with extracellular soluble proteins, and thereby inhibit their growth. Since they have anti-inflammatory and analgesic effects, terpenoids found in many plants have been utilized in traditional medicine [55]. According to Gómez and González [56], terpenes can function as antioxidant compounds through direct ROS scavenging pathways and also modulate the endogenous antioxidant system. Terpenes have demonstrated protective effects against oxidative stress in a variety of diseases, including as cancer, diabetes, liver, renal, neurological, and cardiovascular disorders, as well as in the aging process. This is partially attributed to their antioxidant nature. Their antioxidant action could perhaps account for their ability to control immunological responses, inflammation, and brain signal transmission.

## 6. Saponins

Saponins exhibit special characteristics that distinguish foaming and emulsifying agents [57]. More specifically, they are naturally occurring glycosides that foam like soap when agitated in aqueous solutions, which causes them to create foam. Saponins are triterpenoids with a lipophilic structure and one or more hydrophilic sugar moieties called glycosides [58]. The capacity of precipitated cholesterol to form insoluble complexes is a crucial characteristic of this class of compounds. Both characteristics are linked to amphipathic or amphiphilic traits. Saponins are ubiquitous phytochemicals found in plants that are structurally and functionally diverse. These beneficial bioactive substances play a significant role in plants' defense against pests, diseases, and herbivores due to their antibacterial, antifungal, antiparasitic, insecticidal, and anti-feedant characteristics [59–62]. The bioactivities of saponin are normally assigned to their amphipathic properties, consisting of a hydrophobic triterpene/steroid skeleton and hydrophilic carbohydrate molecules. It is known that the potential biological activities of saponins are dependent on other aspects of their structure.

## 7. Tannins

Tannins are secondary metabolites found in plants that have the power to inhibit insect growth [63]. The term "tannins" refers to a broad class of polymeric phenolic compounds that have the astringency to precipitate gelatin from solution or tan leather. Tannins are anti-nutrients of plant origin as they can inhibit digestive enzymes, precipitate proteins, and reduce the absorption of vitamins and minerals. Epidemiological data suggests that tannin is essentialfor the treatment of skin inflammation, injuries and prevent the onset of chronic diseases [64]. The biological effects of tannin have been extensively investigated using various in vitro or animal models. Nevertheless, there are currently only a few plant extracts with clinical data on humans.

Tannin has a variety of biological activities, such as antimicrobial and antiviral activities, inhibition of lipid peroxidation and lipoxygenases, antioxidant or radical scavenging activity, antimutagenic activity, and antidiabetic qualities [65, 66]. Tannin exhibits antioxidant action due to their ability to scavenge free radicals and reactive oxygen species (ROS) and to chelate transition metal ions, which alter the oxidation process via tannin ortho-dihydroxyl groups [64]. The most important factor in determining the biological activity of tannin is the presence of number of aromatic rings and hydroxyl radicals groups. Therefore, high degree of polymerization and molecular weight of tannins significantly affect their antioxidant activity. Researchers have shown that the higher the relative molecular weight of the substance, the stronger the antioxidant activity of tannin; however, the exact mechanism underlying this activity is yet unknown [67]. The number and degree of polymerization of hydroxyl groups of tannin determine the ability to scavenge free radicals. Hence, the more the hydroxyl groups in plant tannin, the more easy they can be oxidized which confers higher antioxidant activity [68, 69]. In the digestive system, gallic tannin breaks down quickly, whereas condensed tannin is more difficult to break down and absorb. Consequently, the overall antioxidant capacity of plant tannin in animals cannot be easily explained.

## 8. Anthocyanins

Anthocyanins are the largest group of water-soluble pigments in the plant kingdom, displaying a variety of colors that range from pink-orange through red to purple and blue hues. The antioxidant activity of anthocyanins mainly depends on their chemical structure: the position and number of the hydroxyl groups as well as presence or absence of conjugated double bonds and electron donors in the structural ring [70, 71]. It also depends on the type and extent of acylation and glycosylation as well as the degree of hydroxylation at the B-ring [72]. The antioxidant capacity of anthocyanidins decreases in the order of Dp > Pt > Mv = Cy > Pn > Pg (Cy = cyanidin, Dp = delphinidin, Mv = malvidin, Pg = pelargonidin, Pn = peonidin, Pt = petunidin) [73]. Additionally, glycosylation lessens their capacity to scavenge free radicals in comparison to their aglycone forms by reducing the capacity to donate hydrogen, chelate metal, and delocalize electrons [74]. The antioxidant activity of anthocyanins decreases with the number of sugar units at the C3 and C5 locations [72]. To summarize, the number of hydroxyl groups in the B-ring increases the antioxidant activity; however, the number of glycosyl groups attached to the A and C rings decreases it; the latter effect is minimized when the glycosides are acylated [75].

### 8.1 Carotenoids

Carotenoids are powerful scavengers of ROS and highly effective physical and chemical quenchers of singlet oxygen (1O2) [76]. They also provide protection against a number of ROS-mediated disorders, including neurological, cardiovascular, and photosensitive or ocular diseases [77]. The amount of conjugated double bonds in the molecule determines the quenching activity of carotenoids to a greater or lesser degree than the cyclic or acyclic end groups or the type of substituents in those with cyclic end groups. Among the plant-based carotenoids, lycopene (consisting of two non-conjugated and eleven conjugated double bonds) is one of the most effective singlet oxygen quenchers. The antioxidant properties of carotenoids depend on the oxygen tension and concentration present in the tissue [78]. At low oxygen partial pressures, such as those found in most tissues under normal physiological conditions, it inhibits oxidation, whereas at high oxygen tensions, it readily autoxidizes and exhibits pro-oxidant behavior (induces oxidative stress by producing ROS or releasing ATP) [79]. The hydroxy-carotenoids are better radical-trapping antioxidants than β-carotene (due to the presence of hydroxy-functional groups on the ring structures) [80].

Epidemiological studies have claimed that people with high dietary intakes of carotene or high blood levels of this nutrient have lowered risk of various illnesses such as cancer and heart disease [81]. According to Sies and Stahl [82], carotenoids are known to have the chemical capacity to quench singlet oxygen and prevent peroxyl free radical processes. In addition to antioxidant property, carotenoids play a significant role in facilitating cell-to-cell communications via gap junctions. A significant role for dietary carotenoids in protecting against cancer could be due to preventing gap junction communications, which are hindered by many carcinogens [83].

### 8.2 Alkaloids

Alkaloids are the secondary metabolites of plants with lower molecular weight, containing nitrogen atoms, and are derived from amino acids. They are known to protect the plant against herbivores and pathogens. Alkaloids are typically colorless compounds that exhibit optical activity [1]. Alkaloids are fundamental natural products found mostly in various plants and are considered to be some of the most effective and important plant compounds for therapeutic purposes [84].

Many studies have found that alkaloids play a significant role in the prevention of various illnesses including diabetes, cancer, cardiac dysfunction, respiratory disorders, and others, and they are also used as local anesthesia and pain relief [85]. Alkaloids such as cocaine, quinidine, vinblastine, quinine, berberine, and ametrine are used as anesthetic agents, antitumor agents, antimalarials, and antibacterial agents. The most potent alkaloids are those with hydroxyl substitution and a partly desaturated pyridine ring; these alkaloids can quickly donate a hydrogen atom to radicals, scavenging reactive oxygen species (ROS) and RNS [86]. In Sun et al. [87] stated that among the alkaloids found in plants, berberineis the most prominent alkaloids which accounts for approximately 50% of the total alkaloid fraction, while jatrorrhizine, palmatine, and coptisine are the most abundant minor alkaloids. Alkaloids increase superoxide dismutase activity and decrease superoxide anion and malondialdehyde (MDA) formation, as well as bind with catalyzing metal ions (transition metals such as iron and copper ions), which can reduce metal ion concentrations in lipid peroxidation [88]. Apart from these, alkaloids found in various plants and herbs have been shown to have antihypertensive properties. Alkaloids like caffeine, present in several beverages such as tea and coffee, are reported to lower both systolic and diastolic blood pressures. Alkaloids such as colchicine are best known for their preventive action against gout, and they also reduce pain and swelling in degenerative and immunological inflammatory diseases [89].

## 9. Ascorbic acids (vitamin C)

Vitamin C is well known for its antioxidant properties and helps inhibit infection and toxicity [90]. Many studies have also reported that vitamin C plays a significant role in the prevention and treatment of diabetes mellitus, cancer, asthma, cataracts, and HIV infection [91–93]. Vitamin C is a potent antioxidant, as it can donate a hydrogen atom and form a relatively stable ascorbyl free radical. According to Weber et al. [94], ascorbate functions as a scavenger of reactive oxygen species (ROS) and is effective against hydrogen peroxide, singlet oxygen, superoxide radical anion, and hydroxyl radical by accepting hydrogen atoms. Additionally, ascorbyl free radical can be transformed back into reduced ascorbate by absorbing another hydrogen atom. Moreover, vitamin C scavenges reactive nitrogen oxide (RNS) species to stop target molecules from nitrosation. It is necessary to form collagen, an important protein used to make skin, scar tissue, tendons, ligaments, and blood vessels. It is also essential for the repair and maintenance of bones, cartilage, teeth and healing of wounds [95]. It is effective in suppressing the formation of carcinogens such as nitrosamines and protection against oxidative damage [96]. Many research studies have claimed that the intake of vitamin C is inversely related to the incidence of cancer due to its protective effects on the pancreas, lung, stomach, breast, cervix, rectum, and oral cavity [97]. Deficiency of vitamin C causes scurvy, anemia, bleeding gums, infections, nutrient intake in relation to muscle degeneration, poor wound healing, atherosclerotic plaques, and capillary hemorrhaging.

## 10. Tocopherols (vitamin E)

Tocopherol is a well-recognized bioactive compound because of its role in the prevention of ROS generation, thereby preventing the onset of different diseases such as coronary heart disease, arthritis, asthma, cancers, and infectious diseases [98]. It is necessary for the proper functioning of the immune system, cell signaling, and regulation of gene expression, antibody production, phagocytic, and lymphocytic responses, and it exhibits resistance to viral and infectious diseases [98, 99]. They are basically derivatives of the 6-hydroxy chromane ring with three isoprenoid units that form the side chain, and their antioxidant property is due to the presence of the chromane ring [100]. Vitamin E functions primarily as an antioxidant by trapping peroxyl free radicals, and it is considered a master antioxidant as it inhibits bad cholesterol (LDL), which is believed to be the first step in the development of atherosclerosis [101, 102]. Therefore, it helps in preventing or delaying coronary heart disease by limiting the oxidation of LDL-cholesterol and preventing the formation of blood clots, which could lead to a heart attack. Many studies have shown that vitamin E can delay or minimize the development of cataracts, and it also has a protective effect against cataract formation induced by radiation, glucose, or galactose [103]. In vivo research studies have demonstrated the effectiveness of vitamin E supplementation in inhibiting the elevation of free-radical concentrations related to arthritis [104]. Vitamin E has decreased the risk of colon and prostate cancers and also reduced the risk of coronary heart diseases by approximately 40 percent. Several studies have also reported that vitamin E supplementation has been significantly shown to be more effective than placebo in relieving pain [105].

## 11. Evaluation of the antioxidant activity

The ability of bioactive substances to counteract the effects of free radicals is generally linked to their antioxidant activity. Their activity may be primarily based on one of two types of reactions: hydrogen atom transfer (HAT) or electron transfer (ET) [106]. The ability of an antioxidant to reduce radicals, metals, and carbonyl groups is measured by electron transfer assays, whereas the ability to donate hydrogen and so quench free radicals is measured by hydrogen atom transfer assays [107]. The ET assays comprise the following: diphenyl-1-picrylhydrazyl (DPPH) copper (II) reduction capacity, trolox equivalent antioxidant capacity (TRAP), 2,2-Azinobis 3-ethylbenzthiazoline-6-sulfonic acid radical scavenging assay (ABTS+), Folin-Ciocalteu's phenol reagent reducing ability, and ferric ion reducing antioxidant parameter (FRAP). In contrast, the HAT assays comprise the oxygen radical absorbance capacity (ORAC), inhibition of induced low-density lipoprotein, total radical trapping antioxidant parameter (TRAP), LDL oxidation, total radical scavenging capacity assay (TOSCA), β-carotene bleaching, and chemiluminescent assay [108, 109]. However, no single technique can show the effectiveness of the bioactive compounds as an antioxidant, despite research producing a multitude of in vitro assays to assess antioxidant activity. Firstly, the radicals like DPPH and ABTS that are most frequently used in in vitro tests are irrelevant in biological systems. The ones that are relevant are utilized in exorbitant amounts, which once more do not correspond to biological conditions [110]. Additionally, the analysis of the literature demonstrates that results from these assays are not similar between laboratories and are not comparable between two separate antioxidant assays. The reliability of these methods has been called into question and their shortcomings have been highlighted as a result of all of these observations. Cellular assays offer an alternative to these constraints by offering a more accurate assessment of antioxidant activity at the cellular level [111]. These assays mainly comprise the red blood cell hemolysis assay and the more popular human hepatocarcinoma cell cellular antioxidant assay, which offer an affordable intermediate step between in vitro chemical assays and the most trustworthy in vivo investigations. Indeed, in vivo research on humans or animals is unquestionably the most effective method for assessing the antioxidant activity of bioactives. These investigations, however, are costly, time-consuming, intricate, and frequently need for an extensive regulatory procedure [112]. Considering all the aforementioned, the general practice to estimate antioxidant activity by using in vitro assays is currently to use multiple assays (at least three and preferably based on different reaction mechanisms), as well as a reference compound with demonstrated antioxidant activity. It should be possible to find out how bioactive compounds generally have the ability to function as antioxidants with this kind of analysis. Beyond these methodological constraints, behavior and fate of bioactive chemicals in a complex system like the human body pose a significant challenge to their potential as antioxidants.

Although the potential for antioxidant activity can be identified via chemical and cellular assays, the bioaccessibility and bioavailability of bioactives in the human body are the main areas of concern. Bioavailability is defined as the proportion of the compound that can be ingested and utilized while bioaccessibility determines the amount of bioactive compound(s) that can be released from the food matrix into the gastrointestinal tract [113]. Since dietary polyphenols belong to various classes of polyphenols, their bioaccessibility and bioavailability are primarily determined by their structure (polymerization degree, type and quantity of functional groups, etc.) and external factors like interactions with the food matrix and gut microbiota characteristics [114, 115]. The majority of dietary bioactives, particularly polyphenols, are known to have low bioavailability. It is estimated that between 90 and 95% of polyphenols enter the colon intact [116]. At this point, microencapsulation methods are considered to be an effective method for increasing the bioaccessibility and bioavailability of bioactives because they shield the active component from gastrointestinal system degradation, make it more soluble, and promote better penetration of the intestinal barrier [117].

## 12. Antioxidant mechanism

The study of free radical chemistry has received a lot of attention in the last few years. Our body produces reactive oxygen species, reactive nitrogen species, and free radicals when it is exposed to diverse physiochemical circumstances or pathological states, as well as through a variety of endogenous systems. However, for normal physiological function, antioxidants and free radicals must coexist in balance. Oxidative stress occurs when the body cannot control the amount of free radicals in the blood. Free radicals thus adversely alter proteins, lipids, and DNA and trigger a number of human diseases [118]. The process of oxidative metabolism involves energy activation and electron reduction, which leads to the production of reactive oxygen species (ROS) such as superoxide radical anion ($O_2$), singlet oxygen ($^1O_2$), hydrogen peroxide ($H_2O_2$), hydroxyl radical (-OH), and hypochlorous acid (HOCl). The production of reactive oxygen species (ROS) is enhanced during inflammation, radiation exposure, aging, endotoxin shock, and ischemia-reperfusion of the heart, liver, kidney, intestine, and brain. They have been implicated in various cell dysfunctions [119].

## 13. Antioxidant action

The body produces reactive oxygen species (ROS), which are harmful byproducts of aerobic metabolism in active cells. Numerous metabolic routes in living cells produce a large variety of reactive free radicals, which are accounted for by ROS [120]. Under normal physiological circumstances, ROS are regarded as significant regulatory agents in the complicated cell signaling network. They are essential for immunological response, adaptability to metabolic and physiological stressors, cell growth and differentiation, and defense against pathogen invasion [120, 121]. Nonetheless, a number of circumstances may disrupt normal physiological functions, exposing tissues to oxidative stress conditions and leading to an overabundance of reactive oxygen species (ROS) (**Figure 2**). Oxidative stress causes rapid oxidative damage in cells, which sets off a series of degenerative processes. Among the potential

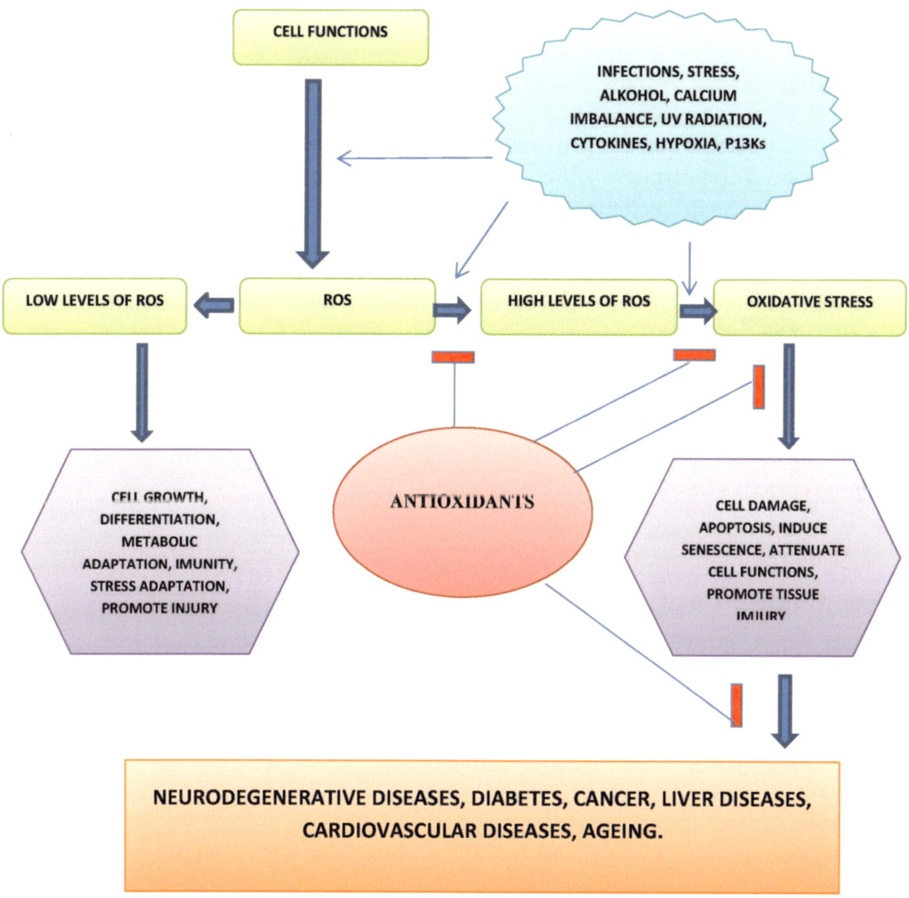


**Figure 2.**
*Reactive oxygen species (ROS) and their contribution to development of oxidative stress-related diseases. The presence of antioxidants may prevent diseases development by modulating damaging effects of ROS by suppressing multiple process steps.*

consequences include the emergence of severe pathologies such as diabetes, cancer, liver illnesses, cardiovascular diseases, neurological diseases, and rapid aging [122, 123]. Several researchers have suggested that the best approach to managing oxidative stress and preventing oxidative damage is to use antioxidants as a therapeutic intervention [124]. In biological systems, a wide range of molecules with different weights and structural characteristics, such as hormones and enzymes, exhibit antioxidant properties [125]. Antioxidants can counteract oxidative stress in a number of ways, such as by preventing the formation of free radicals (preventive antioxidants), stopping autoxidation chain reactions (chain-breaking antioxidants), protecting and enhancing cellular antioxidant defense mechanisms (indirect antioxidants), blockading the action of pro-oxidative enzymes (enzyme inhibitors), neutralizing the action of metal pro-oxidant ions (metal chelators), and boosting the activity of other antioxidants (synergistic compounds) [124, 126]. Numerous studies have shown that phytochemicals, or bioactive compounds, such as tannins, phenolic acids, flavonoids, carotenoids, terpenoids, glycosides, saponins, coumarins, and anthocyanins, are commonly plentiful in medicinal plants. These compounds are important for their antioxidant activity and protect against oxidative damage by scavenging or neutralizing free radicals and reactive oxygen species [127, 128].

## 14. Conclusion

Food bioactive compounds are secondary metabolites that are non-essential or extra-nutritional constituents which typically occur in small quantities in foods, having beneficial health effects. Considering bioactive compounds, as well as antioxidant activities, various plant extracts can be conveniently used as alternatives to synthetic antioxidants. However, scientific validation of plants as modern medicine is lacking. Therefore, folkloric or traditional pharmacological profiles of plants need to be preserved by proper documentation and management. It is advised to focus more research work on the biological and pharmacological activities of bioactive compounds from different plant species.

## Author details

Okram Abemsana Devi* and Mridula Saikia Barooah
Department of Food Science and Nutrition, College of Community Science,
Assam Agricultural University, Jorhat, Assam, India

*Address all correspondence to: okramabemsana@gmail.com

IntechOpen

© 2025 The Author(s). Licensee IntechOpen. This chapter is distributed under the terms of the Creative Commons Attribution License (http://creativecommons.org/licenses/by/4.0), which permits unrestricted use, distribution, and reproduction in any medium, provided the original work is properly cited.

## References

[1] Harborne JR. Introduction to Ecological Biochemistry. Vol. 4. Elsevier, London: Academic Press; 1993. pp. 1-32

[2] Kaur S, Das M. Nutritional and functional characterization of barley flaxseed based functional dry soup mix. Journal of Food Science and Technology. 2014;**52**:5510-5521

[3] Acosta-Estrada BA, Gutiérrez-Uribe JA, Serna-Saldívar SO. Bound phenolics in foods. Food Chemistry. 2014;**152**:46-55

[4] Saponjac VT, Canadanovic-Brunet J, Cetkovi G, Djilas S. Detection of bioactive compounds in plants and food products. In: Emerging and Traditional Technologies for Safe, Healthy and Quality Food. Cham, Switzerland: Springer; 2016. pp. 81-109

[5] Jaime L, Santoyo S. The Health Benefits of the Bioactive Compounds in Foods. Foods. 2021;**10**(2):325. DOI: 10.3390/foods10020325

[6] Guaadaoui A, Benaicha S, Elmajdoub N, Bellaoui M, Hamal A. What is a bioactive compound? A combined definition for a preliminary consensus. International Journal of Nutrition and Food Sciences. 2014;**3**:174. DOI: 10.11648/J.IJNFS.20140303.16

[7] Teodoro AJ. Bioactive compounds of food: Their role in the prevention and treatment of diseases. In: Oxidative Medicine and Cellular Longevity. Hindawi; 2019. pp. 1-4. DOI: 10.1155/2019/3765986. Available from: https://onlinelibrary.wiley.com/doi/10.1155/2019/3765986

[8] Biesalski HK, Dragsted LO, Elmadfa I, Grossklaus R, Muller M, Schrenk D, et al. Bioactive compounds: Definition and assessment of activity. Nutrition. 2009;**25**:1202-1205

[9] Plumb J, Pigat S, Bompola F, Cushen M, Pinchen H, Norby E, et al. eBASIS (bioactive substances in food information systems) and bioactive intakes: Major updates of the bioactive compound composition and beneficial bioeffects database and the development of a probabilistic model to assess intakes in Europe. Nutrients. 2017;**9**:320

[10] Bernhoft A. A brief review on bioactive compounds in plants. Bioactive compounds in plants – benefits and risks for man and animals. The Norwegian Academy of Science and Letters. 2010;**50**:11-17

[11] Jeszka-Skowron M, Zgoła-Grześkowiak A, Grześkowiak T, Ramakrishna A, editors. Analytical methods in the determination of bioactive compounds and elements in food. Food Bioactive Ingredients. 2021. DOI: 10.1007/978-3-030-61879-7

[12] Samtiya M, Aluko RE, Dhewa T, Moreno-Rojas JM. Potential health benefits of plant food-derived bioactive components: An overview. Foods. 2021;**10**(4):839. DOI: 10.3390/foods10040839

[13] Kris-Etherton PM, Hecker KD, Bonanome A, Coval SM, Binkoski AE, Hilpert KF, et al. Bioactive compounds in foods: Their role in the prevention of cardiovascular disease and cancer. The American Journal of Medicine. 2002;**113**(9):71-88

[14] Pillai N, Soumya P, Mini S, Sivan SK, Mondal S. Bioactive compounds in functional food and their role as

therapeutics. Bioactive Compounds in Health and Disease. 2021;**4**(3):24-39. DOI: 10.31989/bchd.v4i3.786

[15] Potter SM, Baum JA, Teng H, Stillman RJ, Shay NF, Erdman JW Jr. Soy protein and isoflavones: Their effects on blood lipids and bone density in postmenopausal women. The American Journal of Clinical Nutrition. 1998;**68**(6):1375S-1379S

[16] Setchell KD, Cassidy A. Dietary isoflavones: Biological effects and relevance to human health. The Journal of Nutrition. 1999;**129**(3):758S-767S

[17] Sandhya VG, Rajamohan T. Beneficial effects of coconut water feeding on lipid metabolism in cholesterol-fed rats. Journal of Medicinal Food. 2006;**9**(3):400-407

[18] Basri DF, Fan SH. The potential of aqueous and acetone extracts of galls of *Quercus infectoria* as antibacterial agents. Indian Journal of Pharmacology. 2005;**37**(1):26-29

[19] Uttara B, Ajay VS, Zamboni P, Mahajan RT. Oxidative stress and neurodegenerative diseases: A review of upstream and downstream antioxidant therapeutic options. Current Neuropharmacology. 2009;**7**(1):65-74

[20] Ramaswamy C. Emerging role of polyphenolic compounds in the treatment of eurodegenerative diseases: A review of their intracellular targets. European Journal of Pharmacology. 2006;**545**(1):51-64

[21] Seifu T. Ethnobotanical and ethnopharmaceutical studies on medicinal plants of Chifra District, Afar Region, North Eastern Ethiopia [MSc thesis]. Addis Ababa, Ethiopia: Addis Ababa University; 2012

[22] Bertschinger HJ. Veterinary reproduction & obstetrics. In: Arthur GH, Noakes DE, Pearson H, Parkinson TJ, editors. Book Review. Journal of the South African Veterinary Association. 7th ed. 1997. p. 68(1). DOI: 10.4102/jsava.v68i1.863

[23] Andjelkovic M, Van Camp J, De Meulenaer B, Depaemelaere G, Socaciu C, Verloo M, et al. Iron-chelation properties of phenolic acids bearing catechol and galloyl groups. Food Chemistry. 2006;**98**(1):23-31

[24] Gee JM, Johnson IT. Polyphenolic compounds: Interactions with the gut and implications for human health. Current Medicinal Chemistry. 2001;**11**:1245-1255

[25] Willcox ML, Bodeker G. Traditional herbal medicines for malaria. Clinical Review. British Medical Journal. 2004;**329**:1156-1159

[26] Dai CC, Xie H, Wang XX, Li PD, Zhang TL, Li YL, et al. Intercropping peanut with traditional Chinese medicinal plants improves soil microcosm environment and peanut production in subtropical China. African Journal of Biotechnology. 2006;**8**:3739-3746

[27] Peng X, Liu J, Wang C, Han Z, Shu Y, Li X, et al. Unusual prenylated phenols with antioxidant activities from Ganoderma cochlear. Food Chemistry. 2015;**171**:251-257

[28] Dubick MA, Omaye ST. Evidence for grape, wine and tea polyphenols as modulators of atherosclerosis and ischemic heart disease in humans. Journal of Nutraceuticals, Functional & Medical Foods. 2001;**3**:67-93

[29] Nardini M, Natella F, Scaccini C. Role of dietary polyphenols

in platelet aggregation. A review of the supplementation studies. Platelets. 2007;**18**:224-243

[30] Renaud S, de Lorgeril M. Wine, alcohol, platelets and French paradox for coronary heart disease. Lancet. 1992;**339**:1523-1526

[31] Aviram M, Fuhrman B. Polyphenolic flavonoids inhibit macrophage mediated oxidation of LDL and attenuate atherogenesis. Atherosclerosis. 1998;**137**:45-50

[32] García-Lafuente A, Guillamón E, Villares A, Rostagno MA, Martínez JA. Flavonoids as anti-inflammatory agents: Implications in cancer and cardiovascular disease. Inflammation Research. 2009;**58**(9):537-552. DOI: 10.1007/s00011-009-0037-3. Epub 2009 Apr 21

[33] Khalid S, Shaheen S, Hussain K, Shahid MN, Sarwar S. Pharmacological analysis of obnoxious water weed: *Eichhornia crassipes* (Mart.) Solms. The Journal of Animal and Plant Sciences. 2020;**30**(6):P1465-P1475. DOI: 10.36899/JAPS.2020.6.0167

[34] Ma X, Li H, Dong J, Qian W. Determination of hydrogen peroxide scavenging activity of phenolic acids by employing gold nanoshell sprecursor composites as nanoprobes. Food Chemistry. 2011;**126**:698-704

[35] Kelly EH, Tagliaferro AR, Dennis JB. Flavonoid antioxidants: Chemistry, metabolism and structure-activity relationships. The Journal of Nutritional Biochemistry. 2002;**13**(10):572-584

[36] Iwashina T. The structure and distribution of the flavonoids in plants. Journal of Plant Research. 2000;**113**:287-299. DOI: 10.1007/PL00013940

[37] Pichersky E, Gang DR. Genetics and biochemistry of secondary metabolites in plants: An evolutionary perspective. Trends in Plant Science Perspectives. 2000;**5**:439-445. DOI: 10.1016/S1360-1385(00)01741-6

[38] Samanta A, Das G, Das S. Roles of flavonoids in plants. International Journal of Pharmaceutical Science and Technology. 2011;**6**:12-35

[39] Azizah M, Pripdeevech P, Thongkongkaew T, Mahidol C, Ruchirawat S, Kittakoop P. UHPLC-ESI-QTOF-MS/MS-based molecular networking guided isolation and dereplication of antibacterial and antifungal constituents of ventilago denticulate. Antibiotics. 2020;**9**:1-34

[40] Yao LH, Jiang YM, Shi JFA, Tomas-Barberan N, Datta R, Singanusong, et al. Flavonoids in food and their health benefits. Plant Foods Human Nutrition. 2004;**59**(3):113-122. DOI: 10.1007/s11130-004-0049-7

[41] Atmani D, Chaher N, Berboucha M, Ayouni K, Lounis H, Boudaoud H, et al. Antioxidant capacity and phenol content of selected Algerian medicinal plants. Food Chemistry. 2009;**112**(2):303-309. DOI: 10.1016/j.foodchem.2008.05.077

[42] Halliwell B. Antioxidants in human health and disease. Annual Review of Nutrition. 1996;**16**:33-50. DOI: 10.1146/annurev.nu.16.070196.000341

[43] Cook NC, Samman S. Flavonoids-chemistry, metabolism, cardioprotective effects, and dietary sources. Journal of Nutritional Biochemistry. 1996;**7**:66-76

[44] Rice-Evans CA, Miller NJ, Bolwell PG, Bramley PM, Pridham JB. The relative antioxidant activities of plant-derived polyphenolic flavonoids. Free Radical Research. 1995;**22**(4):375-383

[45] Pandey A, Pandey RD, Tripathi P, Gupta PP, Haider J, et al. *Moringa Oleifera*

Lam. (Sahijan) - A plant with a plethora of diverse therapeutic benefits: An updated retrospection. Medicinal Aromatic Plants. 2012;**1**:101. DOI: 10.4172/map.1000101

[46] Schroeter H, Boyd C, Spencer JPE, Williams RJ, Cadenas E, Rice-Evans C. MAPK signaling in neurodegeneration: Influences of flavonoids and of nitric oxide. Neurobiology of Aging. 2002;**23**:861-880

[47] Pietta PG. Flavonoids as antioxidants. Journal of Natural Products. 2000;**63**:1035-1042

[48] Hertog MG, Feskens EJ, Hollman PC, Katan MB, Kromhout D. Dietary antioxidant flavonoids and risk of coronary heart disease: the Zutphen Elderly Study. Lancet. 1993;**342**(8878):1007-1011. DOI: 10.1016/0140-6736(93)92876-u

[49] Sharma P, Gujral HS. Cookie making behavior of wheat–barley flour blends and effects on antioxidant properties. LWT - Food Science and Technology. 2014;**55**(1):301-307

[50] Chen X, Li H, Zhang B, Deng Z. The synergistic and antagonistic antioxidant interactions of dietary phytochemical combinations. Critical Reviews in Food Science and Nutrition. 2022;**62**(20):5658-5677. DOI: 10.1080/10408398.2021.1888693

[51] Ludwiczuk A, Skalicka-Wozniak K, Georgiev MI. Terpenoids. Pharmacognosy, Fundamentals, Applications and Strategies; 2017. pp. 233-266. DOI: 10.1016/B978-0-12-802104-0.00011-1

[52] Valko M, Morris H, Cronin MT. Metals, toxicity and oxidative stress. Current Medicinal Chemistry. 2005;**12**(10):1161-1208. DOI: 10.2174/0929867053764635

[53] Agatonovic-Kustrin S, Kettle C, Morton DW. A molecular approach in drug development for Alzheimer's disease. Biomedicine & Pharmacotherapy. 2018;**106**:553-565

[54] Thenmozhi DC, Ramalakshmi V. Preliminary phytochemical screening and antibacterial activity of Pergulariadaemia. International Journal of Pharma and Bio sciences. 2011;**2**:162-166

[55] Hamuel J. Phytochemicals: Extraction Methods, Basic Structures and Mode of Action as Potential Chemotherapeutic Agents, Phytochemicals - A Global Perspective of Their Role in Nutrition and Health. London, UK: InTech; 2012. DOI: 10.5772/26052

[56] Gonzalez-Burgos E, Gomez-Serranillos MP. Terpene compounds in nature: a review of their potential antioxidant activity. Current Medicinal Chemistry. 2012;**19**(31):5319-5341. DOI: 10.2174/092986712803833335

[57] Kregiel D, Joanna B, Izabela W, Hubert A, Charalampos P, Mirko B, et al. Saponin-Based, Biological-Active Surfactants from Plants. Application and Characterization of Surfactants. InTech; 2017. DOI: 10.5772/68062

[58] Vasudeva RN, Sukhendu BG, Pushpalatha B. Triterpenoidsaponins: Are view on biosynthesis, applications and mechanism of their action. International Journal of Pharmacy and Pharmaceutical Sciences. 2015;**7**(1):24-28

[59] Augustin JM, Kuzina V, Andersen SB, Bak S. Molecular activities, biosynthesis and evolution of triterpenoidsaponins. Phytochemistry. 2011;**72**:435-457

[60] Morrissey JP, Osbourn AE. Fungal resistance to plant antibiotics as a mechanism of pathogenesis.

Microbiology and Molecular Biology Reviews. 1999;**63**:708-724

[61] Osbourn A, Goss RJ, Field RA. The saponins: Polar isoprenoids with important and diverse biological activities. Natural Product Reports. 2011;**28**:1261-1268

[62] Sparg SG, Light ME, Van Staden J. Biological activities and distribution of plant saponins. Journal of Ethnopharmacology. 2004;**94**:219-243

[63] Cowan MM. Plant products as antimicrobial agents. Clinical Microbiology Reviews. 1999;**12**:564-582

[64] Serrano J, Puupponen-Pimi R, Dauer A, Aura AM, Saura-Calixto F. Tannins: Current knowledge of food sources, intake, bioavailability and biological effects. Molecular Nutrition & Food Research. 2009;**53**:310-329

[65] Carlsen MH, Halvorsen BL, Holte K, Bohn SK, Dragland S, Sampson L, et al. The antioxidant content of more than 3100 foods, beverages, spices, herbs and supplements used worldwide. Nutrition Journal. 2010;**9**(article nr. 3):1-11

[66] Dolara P, Luceri C, De Filippo C, Femia AP, Giovannelli L, Carderni G, et al. Red wine polyphenols influence carcinogenesis, intestinal microflora, oxidative damage and gene expression profiles of colonic mucosa in F344 rats. Mutation Research. 2005;**591**:237-246

[67] Gülçin L, Huyut Z, Elmasta M, Aboul-Enein HY. Radical scavenging and antioxidant activity of tannic acid. Arabian Journal of Chemistry. 2010;**3**:43-53

[68] Olejar KJ, Ray S, Kilmartin PA. Enhanced antioxidant activity of polyolefin films integrated with grape tannins. Journal of the Science of Food and Agriculture. 2016;**96**:2825-2831

[69] Toshiaki A, Koshiyama I, Fukushima D. Antioxidative properties of procyanidins B-1 and B-3 from Azuki beans in aqueous systems. Agricultural and Biological Chemistry. 1988;**52**:2717-2722

[70] Ali HM, Almagribi W, Al-Rashidi MN. Antiradical and reductant activities of anthocyanidins and anthocyanins, structure-activity relationship and synthesis. Food Chemistry. 2016;**194**:1275-1282

[71] Martín BJ, Sáez-Plaza P, Ramos Escudero F, Jímenez AM, Fett R, Asuero AG. Analysis and antioxidant capacity of anthocyanin pigments. Part II: Chemical structure, color, and intake of anthocyanins. Critical Reviews in Analytical Chemistry. 2012;**42**(2):126-151

[72] Sadilova E, Stintzing FC, Carle R. Anthocyanins, colour and antioxidant properties of eggplant (*Solanummelongena* L.) and violet pepper (*Capsicum annuum* L.) peel extracts. Zeitschrift für Naturforschung C. 2006;**61**:527-535

[73] Pojer E, Mattivi F, Johnson D, Stockley CS. The case for anthocyanin consumption to promote human health: A review. Comprehensive Reviews in Food Science and Food Safety. 2013;**12**:483-508

[74] Zhao CL, Chen ZJ, Bai XS, Ding C, Long TJ, Wei FG. Structure activity relationships of anthocyanidin glycosylation. Molecular Diversity. 2014;**18**:687-700

[75] Liu Y, Tikunov Y, Schouten RE, Marcelis LFM, Visser RGF. Anthocyanin biosynthesis and degradation

mechanisms in Solanaceous vegetables: A review. Frontiers in Chemistry. 2018;**6**:52. DOI: 10.3389/fchem.2018.00052

[76] Fiedor J, Burda K. Potential role of carotenoids as antioxidants in human health and disease. Nutrients. 2014;**6**(2):466-488. DOI: 10.3390/nu6020466

[77] Fiedor J, Fiedor L, Haessner R, Scheer H. Cyclic endoperoxides of β-carotene, potential pro-oxidants, as products of chemical quenching of singlet oxygen. Biochimica et Biophysica Acta. 2005;**1709**:1-4

[78] Palozza P. Prooxidant actions of carotenoids in biologic systems. Nutrition Reviews. 1998;**56**(9): 257-265. DOI: 10.1111/j.1753-4887.1998.tb01762.x

[79] Burton CS. Warts and all, North Carolina. Medizinhistorisches Journal. 1984;**45**:344

[80] Terao J. Antioxidant activity of beta-carotene related carotenoids in solution. Lipids. 1989;**24**:659-669

[81] Poppel VG, Goldbohm RA. Epidemiologic evidence for beta-carotene and cancer prevention. The American Journal of Clinical Nutrition. 1995;**62** (6 Suppl.):1393S-1402S. DOI: 10.1093/ajcn/62.6.1393S

[82] Sies H, Stahl W. Vitamins E and C, beta-carotene, and other carotenoids as antioxidants. The American Journal of Clinical Nutrition. 1995;**62**(6):1315S-1321S. DOI: 10.1093/ajcn/62.6.1315S

[83] Acevedo P, Bertram JS. Liarozole potentiates the cancer chemopreventive activity of and the up-regulation of gap junction communication and connexin 43 expression by retinoic acid and beta-carotene in 10T1/2 cells. Carcinogenesis. 1995;**16**:2215-2222

[84] Okwu DE. Phytochemicals, vitamins and mineral contents of two Nigeria medicinal plants. International Journal of Molecular Medicine and Advance Sciences. 2005;**1**(4):375-381

[85] Eleazu CO, Eleazu KC, Awa E, Chukwuma SC. Comparative study of the phytochemical composition of the leaves of five Nigerian medicinal plants. Journal of Biotechnology and Pharmaceutical Research. 2012;**3**(2):42-46

[86] Nash RJ. Bioactive compounds from natural sources: Isolation, characterisation and biological properties. Phytochemistry. 2001;**57**(5):797-798. DOI: 10.1016/S0031-9422(01)00100-5

[87] Sun L, Gao YH, Tian DK, Zheng JP, Zhu CY, Ke Y, et al. Inflammation of different tissues in spontaneously hypertensive rats. Acta Physiologica Sinica. 2006;**58**:318-323

[88] Mali MC, Harsh N. Nutritional value estimation of the leaves and seeds of Solanumsurattense. Journal of Medicinal Plants Studies. 2015;**3**(1):27-29

[89] Malkinson FD. Colchicine: New uses of an old, old drug. Archives of Dermatology. 1982;**11**:453-457

[90] Datta S, Sinha BK, Bhattacharjee S, Seal T. Nutritional composition, mineral content, antioxidant activity and quantitative estimation of water-soluble vitamins and phenolics by RP-HPLC in some lesser-used wild edible plants. Heliyo. 2019;**5**(3):e01431

[91] Evans RCA, Miller NJ, Paganga G. Structure-antioxidant activity relationships of flavonoids and phenolic acids. Free Radical Biology & Medicine. 1996;**20**(7):933-956

[92] Jaruga P, Jaruga B, Gackowski D, Olczak A, Halota W, Pawlowska M, et al. Supplementation with antioxidant vitamins prevents oxidative modification of DNA in lymphocytes of HIV-infected patients. Free Radical Biology and Medicine. 2002;**32**:414-420

[93] Polidori MC, Stahl W, Eichler O, Niestroj I, Sies H. Profiles of antioxidants inhuman plasma. Free Radical Biology and Medicine. 2001;**30**:456-462

[94] Weber H, Chelatat A, Reymond P, Farmer EE. Selective and powerful stress gene expression in Arabidopsis in response to malondialdehyde. The Plant Journal. 1996;**37**:877-888

[95] Bamji MS. Early nutrition and health-Indian perspective. Current Science. 2003;**85**(8):1137-1142

[96] Gorton HC, Jarvis K. The effectiveness of vitamin C in preventing and relieving the symptoms of virus-induced respiratory infections. Journal of Manipulative Physiological Therapeutics. 1999;**22**(8):530-533

[97] Renaud S, de Lorgeril M. Wine, alcohol, platelets, and the French paradox for coronary heart disease. Lancet. 1992;**339**(8808):1523 1526. DOI: 10.1016/0140-6736(92)91277-f

[98] Tang A, Graham N, Semba R, Saah A. Association between serum vitamin a and E levels and HIV-1 disease progression. AIDS. 1997;**11**:613-620

[99] Allard RW. Genetic changes associated with the evolution of adaptedness in cultivated plants and their wild progenitors. The Journal of Heredity. 1998;**79**:225-238

[100] Evans WJ. Vitamin E, vitamin C, and exercise. The American Journal of Clinical Nutrition. 2000;**72**(2 Suppl.):647S-652S. DOI: 10.1093/ajcn/72.2.647S

[101] Combs GF. The Vitamins. Fundamental Aspects in Nutrition and Health. USA: Academic Press, Inc.; 1992. p. 63

[102] Papas AM. Determination of antioxidant status in humans. In: Papas AM, editor. Antioxidant Status, Diet, Nutrition and Health. London: CRC Press; 1999. pp. 21-35

[103] Ross WM, Creighton MO, Inch WR, Trevithick JR. Radiation cataract formation diminished by vitamin E in rat lenses in vitro. Experimental Eye Research. 1983;**36**:645-653

[104] Yoshikawa T, Tanaka H, Kondo M. Effect of vitamin E on adjuvant arthritis in rats. Biochemical Medicine. 1983;**29**:227-234

[105] Machtey I, Ouaknine L. Tocopherol in osteoarthritis: A controlled pilot study. Journal of the American Geriatrics Society. 1978;**26**:328-330

[106] Prior RL. Oxygen radical absorbance capacity (ORAC): New horizons in relating dietary antioxidants/bioactives and health benefits. Journal of Functional Foods. 2015;**18**:797-810

[107] Prior RL, Wu X, Schaich K. Standardized methods for the determination of antioxidant capacity and phenolics in foods and dietary supplements. Journal of Agricultural and Food Chemistry. 2005;**53**:4290-4302

[108] Gulcin I. Antioxidants and antioxidant methods: An updated overview. Archives of Toxicology. 2020;**94**:651-715

[109] Huang D, Boxin OU, Prior RL. The chemistry behind antioxidant capacity

assays. Journal of Agricultural and Food Chemistry. 2005;**53**:1841-1856

[110] Apak R. Current Issues in Antioxidant Measurement. Journal of Agricultural and Food Chemistry. 2019;**67**:9187-9202

[111] Shahidi F, Zhong Y. Measurement of antioxidant activity. Journal of Functional Foods. 2015;**18**:757-781

[112] Granato D, Shahidi F, Wrolstad R, Kilmartin P, Melton LD, Hidalgo FJ, et al. Antioxidant activity, total phenolics and flavonoids contents: Should we ban in vitro screening methods? Food Chemistry. 2018;**30**(264):471-475. DOI: 10.1016/j.foodchem.2018.04.012

[113] Carbonell-Capella JM, Buniowska M, Barba FJ, Esteve MJ, Frígola A. Analytical methods for determining bioavailability and bioaccessibility of bioactive compounds from fruits and vegetables: A review. Comprehensive Reviews in Food Science and Food Safety. 2014;**13**:155-171

[114] Kasote DM, Jayaprakasha GK, Patil BS. Encapsulation of polyphenols: An effective way to enhance their bioavailability for gut health. Advances in Plant Phenolics: From Chemistry to Human Health. 2018:239-259. DOI: 10.1021/bk-2018-1286.ch013

[115] Shahidi F, Peng H. Bioaccessibility and bioavailability of phenolic compounds. Journal of Food Bioactives. 2018;**4**:11-68

[116] Etxeberria U, Arias N, Boqué N, Macarulla MT, Portillo MP, Martínez JA, et al. Reshaping faecal gut microbiota composition by the intake of trans-resveratrol and quercetin in high-fat sucrose diet-fed rats. The Journal of Nutritional Biochemistry. 2015;**26**:651-660

[117] Gowd V, Bao T, Wang L, Huang Y, Chen S, Zheng X, et al. Antioxidant and antidiabetic activity of blackberry after gastrointestinal digestion and human gut microbiota fermentation. Food Chemistry. 2018;**269**:618-627

[118] Lobo JM, Valverde JA, Hortal J. The uncertain nature of absences and their importance in species distribution modelling. Ecography. 2010;**33**:103-114

[119] Opie L, Smith A, Spalding J. Conscious appreciation of the effects produced by independent changes of ventilation volume and of end-tidal PCO2 in paralysedpatients. The Journal of Physiology. 1989;**149**:494-499

[120] Pappa HM, Gordon CM, Saslowsky TM. Vitamin D status in children and young adults with inflammatory bowel disease. Pediatrics. 2006;**118**(5):1950-1961

[121] Jahnsen J, Falch JA, Mowinckel P, Aadland E. Vitamin D status, parathyroid hormone and bone mineral density in patients with inflammatory bowel disease. Scandinavian Journal of Gastroenterology. 2002;**37**(2):192-199

[122] Harries AD, Brown R, Heatley RV, Williams LA, Woodhead S, Rhodes J. Vitamin D status in Crohn's disease: Association with nutrition and disease activity. Gut. 1985;**26**(11):1197-1203. DOI: 10.1136/gut.26.11.1197

[123] Vaidya ADB. Reverse pharmacological correlates with ayurvedic drug actions. Indian Journal of Pharmacology. 2008;**38**:311-315

[124] Froicu M, Weaver V, Wynn TA, McDowell MA, Welsh JE, Cantorna MT. A crucial role for the vitamin D receptor in experimental inflammatory bowel diseases. Molecular Endocrinology. 2003;**17**(12):2386-2392

[125] Yu J, Ahmedna M, Goktepe I. Effects of processing methods and extraction solvents on concentration and antioxidant activity of peanut skin phenolics. Food Chemistry. 2005;**90**:199-206

[126] Church TS, Earnest CP, Wood KA, Kampert JB. Reduction of C-reactive protein levels through use of a multivitamin. The American Journal of Medicine. 2003;**115**(9):702-707

[127] Ali BH, Blunden G, Tanira MO, Nemmar A. Some phytochemical, pharmacological and toxicological properties of ginger (*Zingiberofficinale roscoe*): A review of recent research. Food and Chemical Toxicology. 2008;**46**:409-420

[128] Packer L, Rimbach G, Virgili F. Antioxidant activity and biologic properties of aprocyanidin-rich extract from pine (*Pinusmaritima*) bark, pycnogenol. Free Radical Biology & Medicine. 1999;**27**:704-724

**Chapter 5**

# Phytochemicals from Secondary Metabolism and Their Role as Antioxidant and Anti-Inflammatory Molecules

*Ana Montserrat Corona-España, Mario Alberto Garcia-Ramirez, Regina Romo-Gonzalez, Ingrid Mayanin Rodriguez-Buenfil and Orfil González Reynoso*

## Abstract

Inflammation is the natural response to several kinds of stimulus that activate the famous five stages: redness, heat, pain, swelling, and impaired function. This process involves the cytokines release, immune cells action, and molecules transport. Phytochemicals such as flavonoids, polyphenols, and tannins are usually obtained from plants and feature anti-inflammatory properties. Those molecules inhibit pro-inflammatory enzymes as well as modulate cellular signaling, by reducing oxidative stress. Recent researches highlight thiosulfinates and polysulfides as a potential anti-inflammatory option by suppressing NF-kB pathway and promoting tissue's health. This chapter delves into biochemical pathways generating inflammation, phytochemicals compounds, as well as highlighting the crucial role that these phytochemicals play on inflammation modulation and oxidative stress reduction.

**Keywords:** inflammation, biochemical pathways, phytochemicals, antioxidant molecules, anti-inflammatory molecules

## 1. Introduction

Inflammation is a complex metabolic process that appears as a natural response due to an external stimulus such as injury, infection, or irritation. It involves several biological processes that encompass inmmune cells and signaling molecules as a tissue response to eliminate harmful stimulus for the initialization of the healing process [1]. In order to alleviate inflammation and support health, herbal traditional medicine has been used worldwide since ancient times. Plants have been recognized for the key therapeutic properties they present, including the ability to modulate inflamma-tion and oxidative stress. Selected features of plant compounds such as flavonoids, polyphenols, and terpenoids possess antioxidant and anti-inflammatory properties.

These bioactive compounds can scavenge free radicals, reduce oxidative damage, and regulate inflammatory pathways on the body, for example, turmeric, ginger, and green tea [2, 3].

Nowadays, recent research focused on medicinal properties of plants in order to uncover new compounds and action mechanisms that exert therapeutic effects. Either incorporating those plant-based remedies into diets or as supplements, by offering a natural alternative with less side effects, compared to conventional anti-inflammatory drugs. Before introducing ourselves to phytochemicals and their contributions to health, it is important to understand the phenomenon of inflammation.

## 2. Inflammation

Inflammation is the nonspecific immune response against physical, chemical, or even biological attack to the body [4]. It does result in a series of physiological changes that provide the indistinguishable redness characteristics, heat in the affected area due to an increase in blood flow and vascular permeability, pain, swelling, and movement difficulty or loss of function, produced by mediating substances prorelease, immune inflammation cells transport and molecules [1]. Along the process, numerous molecules play an important role. Once the harmful stimulus exists, the release of proinflammatory proteins, known as cytokines, begins. Those are responsible for the NF-kB activation, also called nuclear factor kappa B. NF-kB is a proteins group that control several cellular functions that goes from growth to cell survival, modulating immune and inflammatory responses. It can be found in higher-than-normal amounts in some cells, being indicative of inflammatory disorders (asthma and ulcerative colitis), as well as autoimmune disorders (rheumatoid arthritis) [3]. On the other hand, signaling proteins like cytokines regulate the communication between immune system cells. Playing a fundamental role in the inflammatory response and immunity. During cytokines production, NF-kB regulates the genes expression involved in different cell types, such as macrophages, dendritic cells, and endothelial cells [5]. The NF-kB activation initiates the expression of proinflammatory cytokines, for example, tumor necrosis factor alpha (TNF-a), interleukin-1 (IL-1), interleukin-6 (IL-6), and interleukin-8 (IL-8), among others, helping on the immune cells mobilization towards infection sites or tissue damage [6].

When talking about inflammation, it is important to pantout the activation of NF-kB and the existing relationship with cytokines. Nevertheless, there are other types of molecules such as modified amino acids, neurotransmitters, and inflammatory mediators that participate in the inflammatory mediators release [3]. Amino acids feature alterations either chemically or structurally in contrast to the standard amino acids present in proteins, and those are known as modified amino acids. For example, Citrulline, ornithine, and 5-hydroxytryptophan (5-HTP) are shown in **Figure 1**. Those are some modified amino acids that play a key role on the inflammatory mediators release [3, 8].

Histamine as well as serotonin are neurotransmitters and chemical messengers that serve on significant functions within the nervous system by having a distinct role in contributing to the inflammatory process [8]. Histamine is released by mast cells and basophils in response to allergic or infectious stimulus [6, 8]. It has pro-inflammatory effects by increasing vascular permeability and causing vasodilation. Moreover, it facilitates the inflammatory cells entry to the inflammation site. Additionally, histamine can stimulate the proinflammatory cytokines production

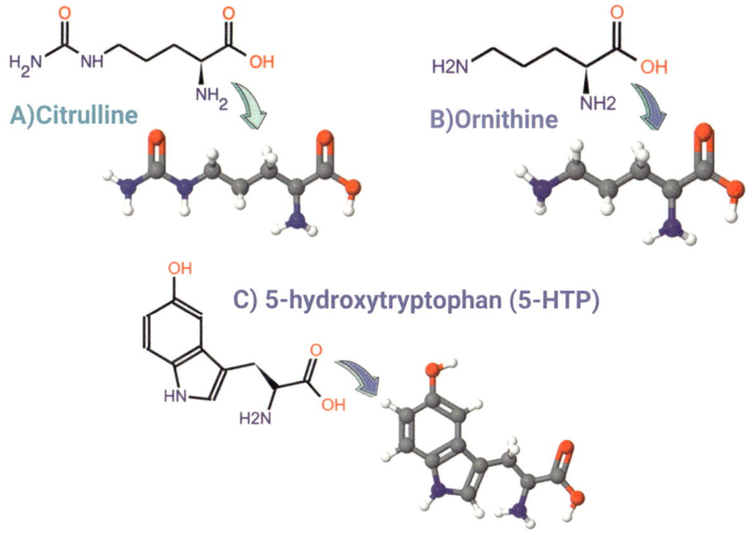


**Figure 1.**
*Examples of modified amino acids that participate in the inflammatory mediators release. Where (a) Citrulline molecule helps improve blood flow and reduce blood pressure, while (b) Ornithine is useful in different metabolic process, and (c) 5-hydroxytryptophan (5-HTP) a chemical byproduct of the protein building block L-tryptophan. Source: PubChem, 2024. Own modification [7].*

such as tumor necrosis factor alpha (TNF-a) and interleukins (IL-1, IL-6) [6]. In contrast, serotonin is primarily a neurotransmitter in the enteric nervous system. It is also synthesized by macrophages and dendritic cells. Studies have demonstrated the pro-inflammatory effects by increasing the inflammatory cells activity and by promoting the pro-inflammatory cytokines secretion. Furthermore, serotonin can modulate the immune response and inflammation through specific receptors on different cell types resulting in chronic inflammation. It can be related to chronic diseases such as inflammatory bowel disease, rheumatoid arthritis, and allergic disorders [3].

Finally, the last group of inflammatory mediators are those derived from arachidonic acid, phospholipases (A2). Phospholipases are responsible for the arachidonic acid conversion into araquinodate. Cyclooxygenases (COX1, COX2), primarily via the COX2 pathway, stimulate the secretion of prostaglandins as PGE2, PGF2, PGD2, PG12, thromboxanes, and TXA2 ultimately leading to the lipoxygenases (LOX) generation responsible for synthesizing leukotrienes [8].

## 3. Oxidative stress

The imbalance between free radicals and their stabilizing agent's antioxidant enzymes is known as oxidative stress. When we talk about oxidative stress, it is important to approach reactive oxygen species or free radicals definition that can be produced by normal cellular metabolism, as well as the interaction with biomolecules such as protein, lipid, and DNA, causing cellular damage and degenerative changes [9]. Free radicals are chemical species that contains unpaired electrons that increase the chemical reactivity of an atom or molecule. Some examples of free radicals include hydroxyl radical (OH), superoxide anion (Oz-), transition metals such as iron and copper, nitric oxide (NO), and peroxynitrite (ONOO) [10].

These molecules are produced through biochemical processes including reduction of molecular oxygen during aerobic respiration, yielding superoxide and hydroxyl radicals, by-products of oxidation of catecholamines, and activation of the arachidonic acid cascade product electrons that can reduce molecular oxygen to superoxide, in addition with the superoxide production and hypochlorous acid (HOC1). Hypochlorous acid is a powerful oxidant, which activates phagocytes as well as nitric oxide production by vascular endothelium cells [9, 10].

However, this reactive species at low concentrations plays a vital role in the physiological regulation and cellular signaling processes, while high level can cause deleterious changes in the cell [9]. Cellular damage occurs on different macromolecules, starting with lipids, which suffer the greatest damage through a process known as lipid peroxidation, that affects structures rich in fatty acids, altering the permeability of the cell membrane, producing edema as well as cell death. The production of free radicals leads to the formation of organic peroxides and other products. Once formed, these free radicals are responsible for cytotoxic effects. On the other hand, in proteins there is an oxidation of amino acids such as phenylalanine, tyrosine, histidine, and methionine that can form cross-linked peptide chains and carbonyl groups. While in deoxyribonucleic acid (DNA) phenomena of mutations and carcinogenesis can take place, due to the loss of expression or synthesis of proteins, caused by the damage of a specific gene, triggered by oxidative modifications of the bases, deletions, fragmentations, stable DNA-protein interactions, chromosomal rearrangements, and demethylation of DNA cytosines that activate genes [11].

To prevent these harmful health effects, the body has a defense against harmful levels of oxidants that consist of different types of antioxidant enzymes that together with their support systems, as well as low molecular mass antioxidants, form an antioxidant network. Antioxidants prevent other molecules from binding to oxygen, by reacting-interacting faster with free oxygen radicals and reactive oxygen species than with the rest of the molecules present. The action of antioxidants could be said that they sacrifice its own molecular integrity to avoid alterations of molecules (lipids, proteins, and DNA) that are functionally vital or more important [11, 12].

There are different types of antioxidant molecules, among it, flavonoids, tannins, polyphenols, polysulfides, and thiosulfinates stand out, which have proven to be good either neutralizing oxygen or capturing free radicals [11].

## 4. Phytochemicals formation through secondary metabolism

Plants, as living organisms, show metabolic processes. Those processes related to photosynthesis, carbohydrates degradation, the nucleic acids production, and proteins formation are vital for the daily processes to ensure the plant survival, also known as the primary metabolism [2]. In contrast, the plants' secondary metabolism represents an alternative for the production and accumulation of compounds of diverse chemical nature. It does facilitate the life of the plant as well as the interaction with its surroundings and environment. As a result of this process, phytochemicals are produced that show biological activity [13].

In **Figure 2**, a general metabolic map is presented. Where primary metabolism is in all processes related to the production or carbohydrates degradation, lipids, nucleic acids, and proteins [13]. The shikimic, mevalonic, and malonic acid pathways are the main processes that make up secondary metabolism, resulting in the phytochemicals formation such as alkaloids, lignins, flavonoids, tannins, flavonoids,

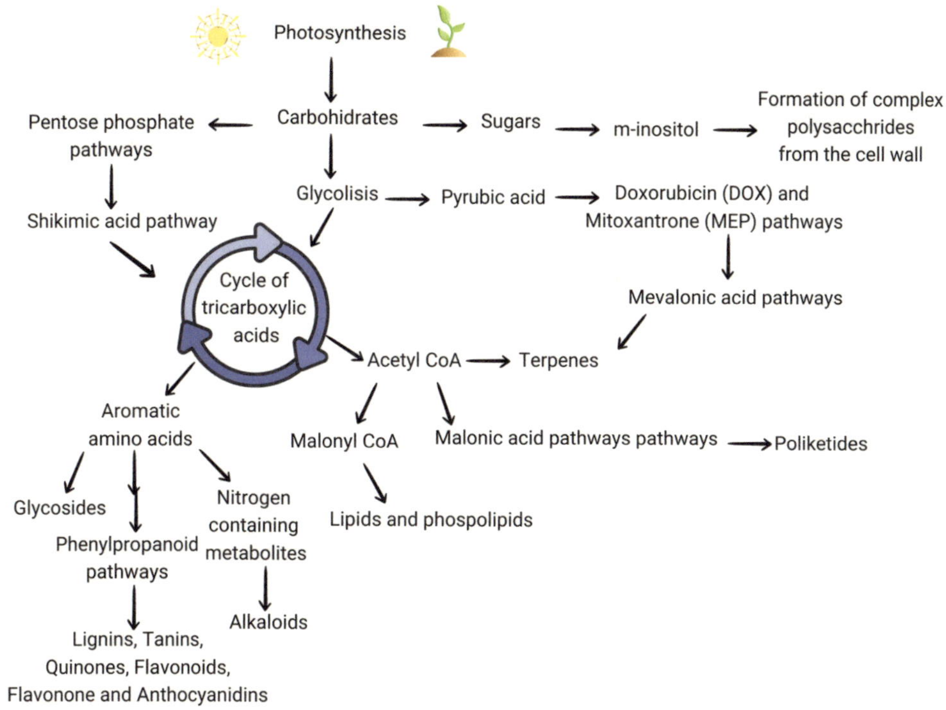


**Figure 2.**
*Plants' metabolic map, for the primary metabolites (sugars, nucleic acids, lipids, and phospholipids) production and secondary metabolites (Terpenes, alkaloids, polyketides, lignin, tannins, flavonones, and anthocyanins). Source: Mendoza and Escamilla, 2018. Own modification [13].*

and anthocyanins, to mention a few of them [2]. After observing this reaction series, it can be appreciated the harmonious and precise coordination that must exist between those processes that occur simultaneously. Phytochemicals have a restricted distribution in the plant kingdom; that means, not all of them are found in all group of plants [13]. Those are synthesized in small quantities and not in a generalized way, often restricting the production to a certain plants genus, a family, or even some species [14].

Secondary metabolites foreseen distinct ecological roles. It works as either attractants or repellents for animals. Among those, certain compounds act as pigments, enhancing the flowers attractiveness and fruits to facilitate pollination, thus contributing significantly to reproduction. On the contrary, others fulfill a defensive function against predators, acting as pesticides or natural detergents [13].

In order to analize the secondary metabolites, as well as the production and uses of these, it is essential to know the cell origin, types, plant tissues, and necessary sugars of such substances in addition with physiological role they play.

Glycolysis and the pentose phosphate pathway are key pathways of carbohydrate degradation in plants. Glycolysis (shown in **Figure 2**) is important to carry out both the tricarboxylic acid cycle and the formation of pyruvic acid, which after a series of transformations and supported by the mevalonic acid pathway lead to the formation of metabolites such as terpenes [15]. On the other hand, the pentose phosphate pathway helps with the formation of the necessary precursors as well as the energy necessary to carry out the shikimic acid pathway, which, in company of the tricarboxylic acid cycle, leads to the aromatic compounds synthesis in the form of glycosides, alkaloids, lignins, tannins, etc. (**Figure 3**) [13, 15].

There are three biochemical pathways responsible for synthesizing those compounds along with three distinct organically identifiable families that are commonly referred as: shikimic, malonate acetate and mevalonic acid. The shikimic acid pathway

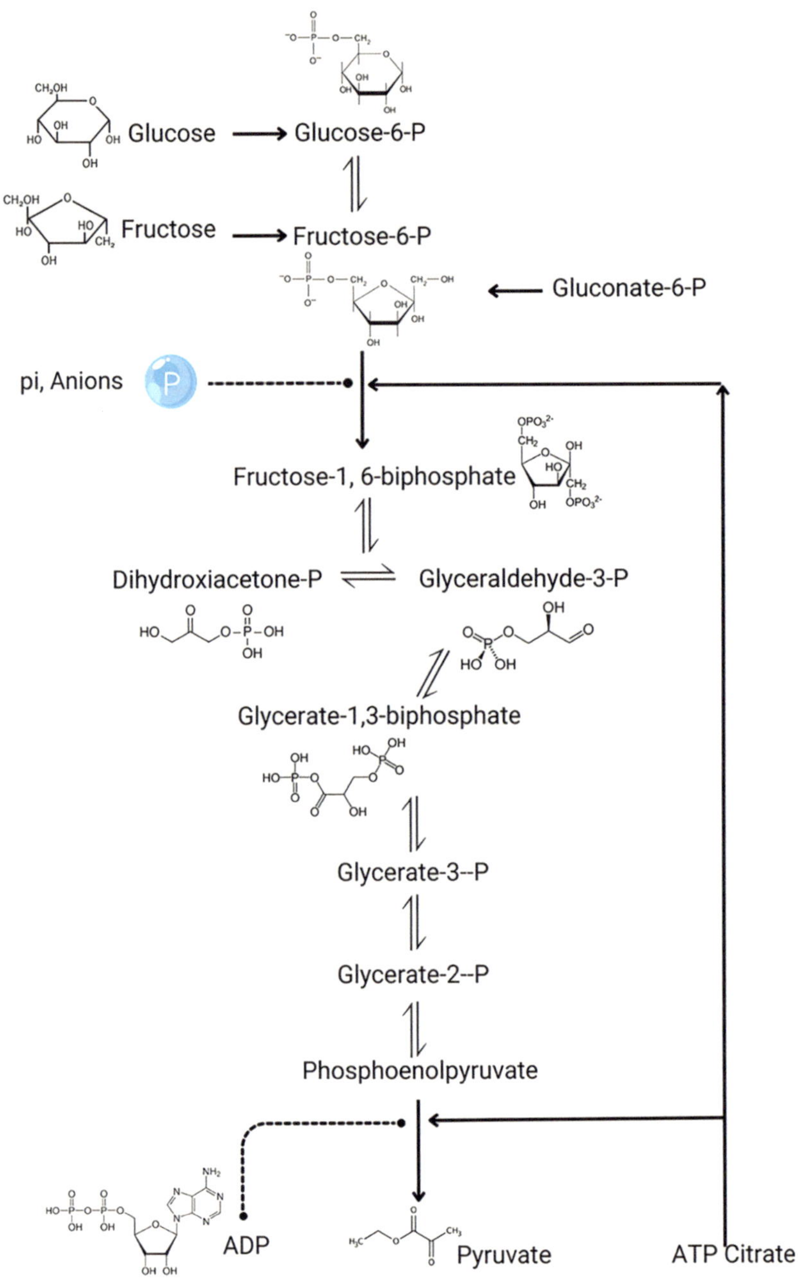


**Figure 3.**
*Glycolysis pathway in plants. Where the action of ADP as a substrate is indicated by a dotted line, presenting a positive effector as well as a negative effector effects, for the formation of pyruvate. Source: Copeland and Turner, 1987. Own modification [15].*

is related to the carbohydrates metabolism and aromatic amino acids. It consists of seven steps, it starts with the phosphoenolpyruvate and erythrose-4-phosphate condensation [16], that interact with the tricarboxylic acid cycle to produce aromatic and aliphatic amino acids such as alkaloids and phenolic compounds as shown in **Figure 4**.

**Figure 5** represents the mevalonic acid pathway. It is associated with the production of secondary metabolites originated from the conjunction of acetyl CoA and

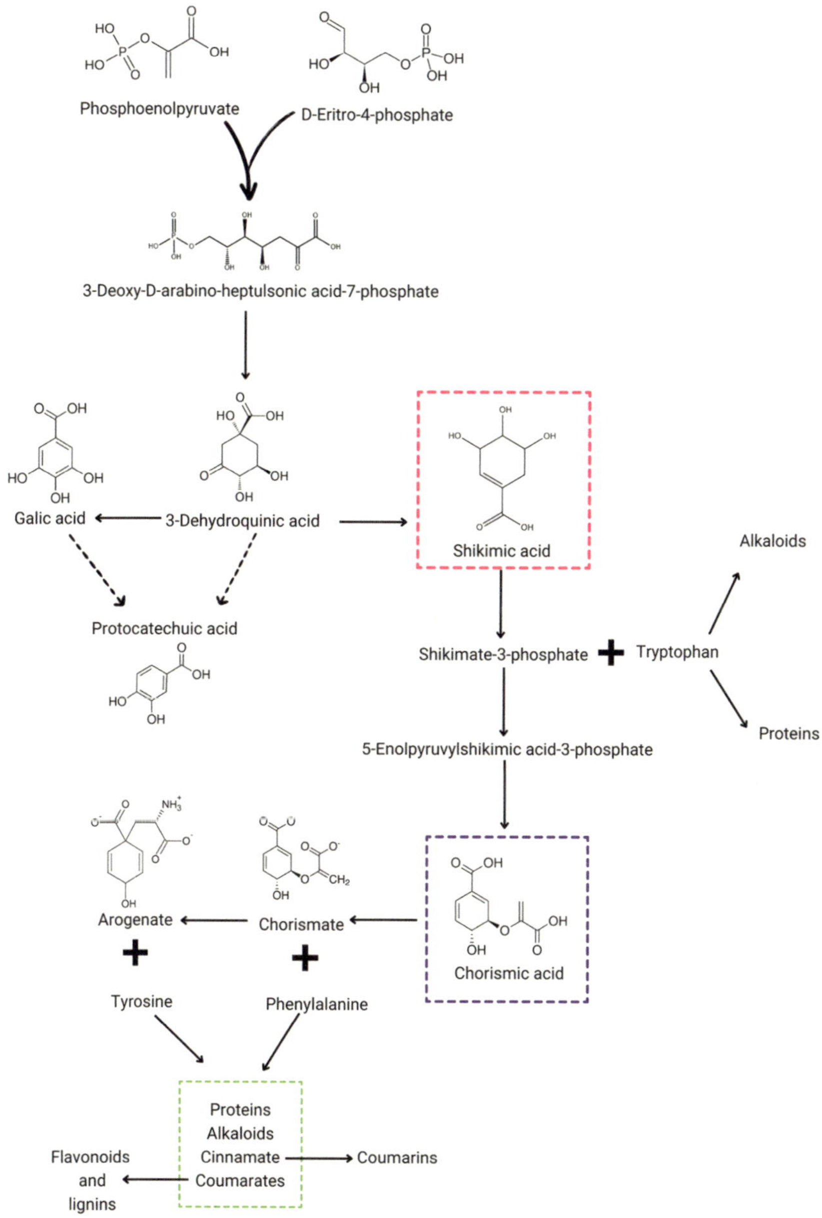


**Figure 4.**
*Shikimic acid pathway, whose objective is the obtention of shikimic acid and chorismic acid, for the formation of phytochemicals such as alkaloids, coumarins, flavonoids, and lignins, from intermediates of the pentose phosphate pathway and the Calvin cycle. Source: Kougan and Medicinal Plant Research in Africa, 2013. Own modification [16].*

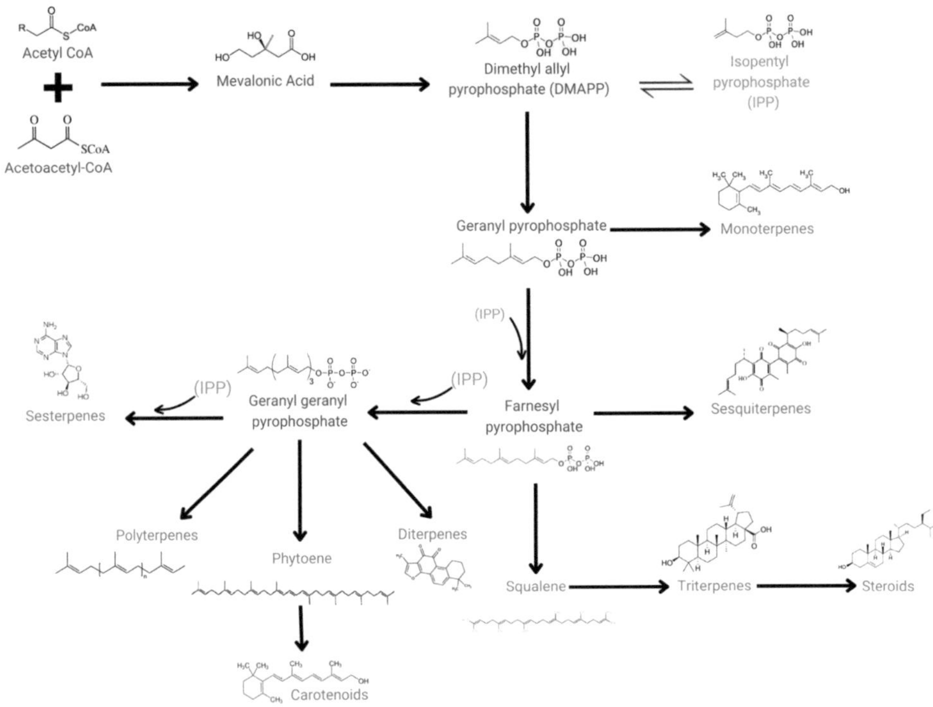


**Figure 5.**
*The union of acetyl CoA and acetoacetyl CoA gives rise to the formation of mevalonic acid that after several transformations and additions of isopentyl pyrophosphate (IPP); it is known as the mevalonic acid pathway, responsible for the formation of terpenes, carotenoids, and steroids. Source: Nissen [17]. Own modification [17].*

acetoacetyl CoA. From the formation of mevalonic acid, it is possible to obtain dimethyl allyl pyrophosphate to get isopentyl pyrophosphate that is decisive for the terpenes formation in different forms (monoterpenes, triterpenes, sesquiterpenes, etc.), carotenoids, and steroids [18].

Whole kind of organisms synthesize isoprenoids and derivatives through the mevalonate (MVA) pathway, as shown in **Figure 6**. It plays a key role in cellular functions communication [20]. The acetate-malonate pathway is also known as the polyketide pathway. It uses shikimate acid derivatives as initial compounds that involve the combination of acetyl CoA via acetate to generate metabolites. Benzophenone synthase enzyme is responsible for producing a variety of secondary metabolites by altering the starter substrate, resulting in the production of phenols and poliphenols as shown in **Figure 6** [21].

Early inhibition of 3-hydroxyl-3-methylglutaryl co-enzyme A reductase (HMGR) in mammals is used to treat hypercholesterolemia [20, 21]. Mutations in mevalonate kinase (MVK) can result in inflammatory diseases such as hyper-immunoglobulin D periodic fever syndrome (HIDS) and mevalonate aciduria (MA) [21]. Conversely, the mevalonic acid pathway in plants is still under investigation, and its significance in growth and adaptation to several stresses induced by pests, pathogens, or environmental changes is well-established [20, 21]. Isoprenoids and terpenes in plants play crucial roles in defense mechanisms against stresses, essential components synthesis such as phytosterols and dolichols as well as the production of volatile compounds for protection and communication [20].

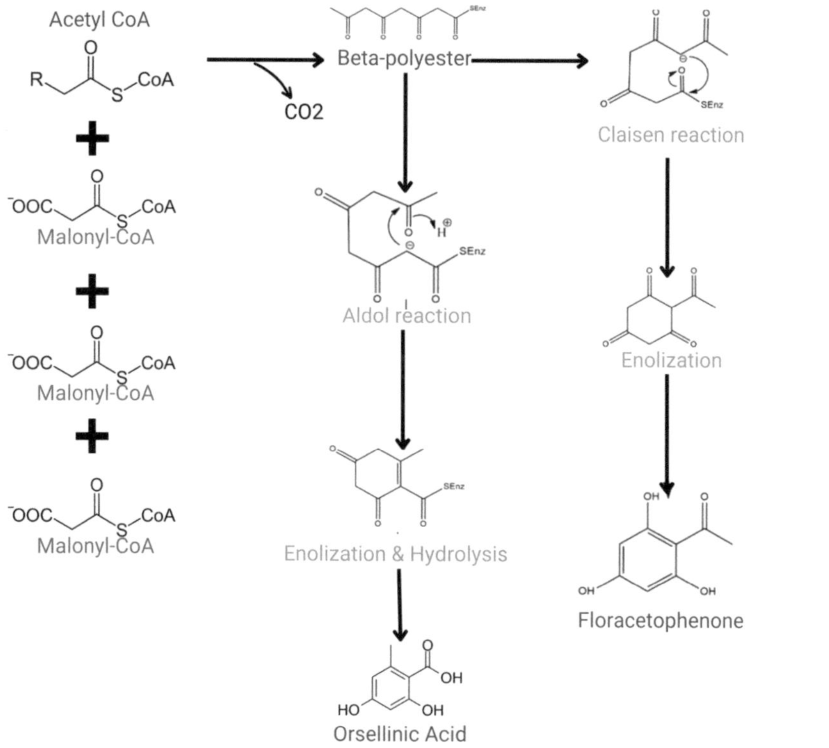


**Figure 6.**
*From acetyl CoA and 3 units of malonyl CoA, betapolyester is formed, giving rise to the orsellinic acid and floracetophenone formation after aldol and Claissen reactions for the synthesis of phenols, in a process known as the acetate–malonate pathway. Source: Dewick, 2009. Own modification [19].*

## 5. Phenols and polyphenols

Secondary metabolites with a fenol group, an aromatic ring with a hydroxyl group or derivade from phenol, are called either phenolic compounds, polyphenols, or phenyl-propanoids, as it can be appreciated in **Figure 7**. Seen from a chemical perspective, phenolic compounds are a diverse group that comprises simple molecules such as phenolic acids, to complex polymers like condensed tannins and lignin. Flavonoid pigments are also found within the phenolic compounds [22].

Phenols are one of the most prominent groups among secondary metabolites in plants and those are produced through several metabolic pathways such as shikimic, acetate–malonate, and acetic acid. The main synthesis occurs through the phenylpropanoid pathway, where the combination of these routes leads to the formation of flavonoids that are the most common phenols in nature. The shikimic acid pathway is crucial for the generation of these compounds and begins with the aromatic amino acids phenylalanine, tyrosine, and tryptophan, derived from phosphoenolpyruvate and erythrose-4-phosphate. The general structure of these molecules is shown in **Figure 7** [23].

Polyphenols present a variety of beneficial effects such as their antimicrobial, anticancer, and anti-inflammatory properties. Where one of their main characteristics is the high antioxidant capacity [24]. The antioxidant and anti-inflammatory properties, along with other polyphenols biological roles, are mainly linked to the specific chemical compositions. These compounds possess aromatic characteristics

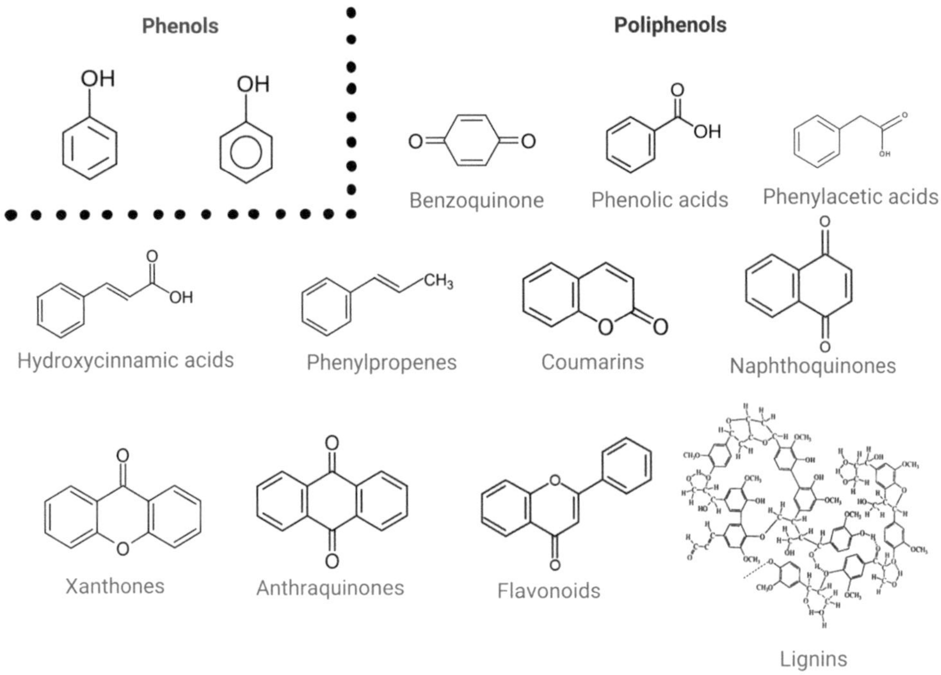


**Figure 7.**
*Phenols and polyphenols chemical structure (benzoquinone, phenolic acids, phenylacetic acid, hydroxycinnamic acids, phenylpropenes, coumarins, naphthoquinones, xanthones, anthraquinones, flavonoids, and lignins). Source: PubChem, 2024. Own modification [7].*

and highly conjugated systems with numerous hydroxyl groups, enabling them to act as effective electrons or hydrogen atoms donors and allowing them to neutralize free radicals and other reactive oxygen species (ROS) [25].

Phenolic compounds can influence cell signaling, suggesting that the anti-inflammatory properties of polyphenols include the inhibition of enzymes such as cyclooxygenase (COX-2), lipoxygenase (LOX), and inducible nitric oxide synthase (iNOS), through the activation of peroxisome proliferator-activated gamma receptor (PPARγ) [26]. Nevertheless, the inhibition of phosphoinositide 3-kinase (PI 3-kinase), tyrosine kinases, nuclear factor-kappa B (NF-kB), c-JUN as well as the activation of phase II antioxidant enzymes, and other kinases leads to the modulation of genes related to survival [25, 26].

A relevant anti-inflammatory mechanism for polyphenols is the inhibition of enzymes that generate eicosanoids such as phospholipase A2. Cyclooxygenase and lipoxygenase reduce the prostanoids and leukotrienes concentration. Arachidonic acid (AA) is released from membrane phospholipids through the action of phospholipase A2. It can then be metabolized by cyclooxygenase (COX) to form prostaglandins (PGs) and thromboxane A2 or by lipoxygenase (LOX) to generate hydroperoxyeicosatetraenoic acids (HpETEs), hydroxyeicosatetraenoic acids (HETEs), and leukotrienes [26].

## 6. Tannins and flavonoids

Within phenols and polyphenols, there are quite a few compounds whose formation is derived from the principles mentioned elsewhere; these compounds are tannins and flavonoids as shown in **Figure 8** [23].

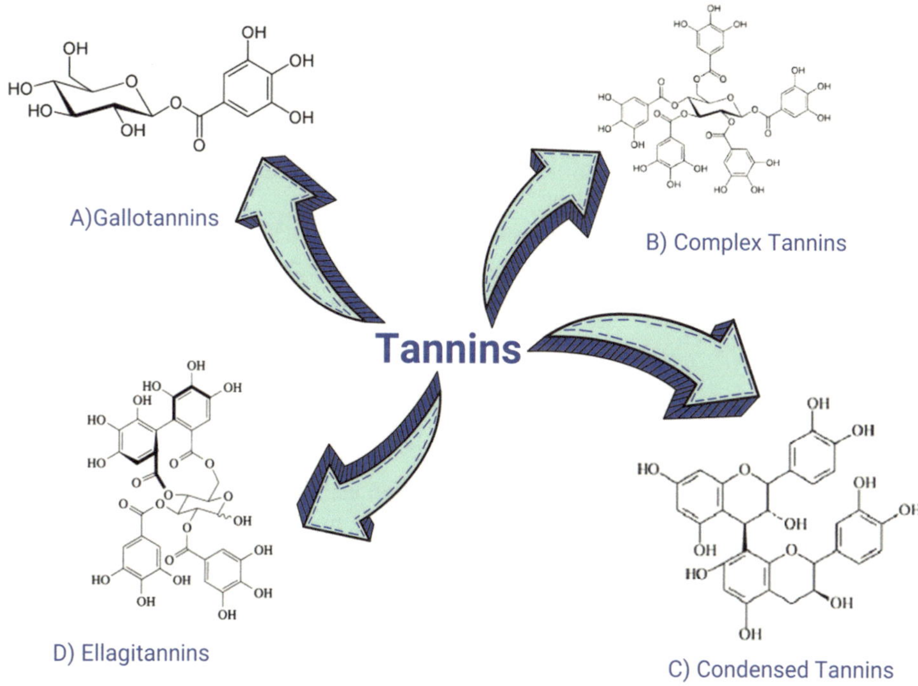


**Figure 8.**
*Tannins chemical structure includes gallotannins, ellagitannins, complex tannins, and condensed tannins. Source: PubChem, 2024. Own modification [7].*

Tannins are a secondary metabolites group from plants that have the ability to tan and convert animal skin into leather. These compounds can be formed from quinic acid and yield gallic acid, obtaining hydrolyzable tannins (HT) biosynthesis as the main product. The formation can occur through a glycosylation of catechin unit to a gallotannin or an ellagitannin unit that subsequently suffers an oligomerization resulting in the formation of a catechin unit which is a condensed tannin [27, 28].

Flavonoids are secondary metabolites of plants with a variable phenolic structure that shows in **Figure 7** and classified depending on the carbon of the C ring. Where the B ring is attached and the unsaturation degree of C ring oxidation. They are abundant in fruits, vegetables, grains, bark, roots, stems, flowers, tea, and wine. Their formation occurs mainly through the shikimic acid pathway [29, 30].

For plants, those compounds have different uses; for example, tannins serve both as a defense mechanism against herbivores and pathogens as well as the nitrogen conservation [31]. While flavonoids provide the plant with characteristics such as color and aroma. In fruits, those attract pollinators. Flavonoids protect from different biotic and abiotic stresses and act as unique UV filters and as a signaling molecule [30].

Tannins and flavonoids have been of great interest due to the multiple therapeutic properties where antioxidant and anti-inflammatory properties stand out. Flavonoids main aim is to be exogenous antioxidants that act by inhibiting the nitric-oxide synthase and xanthine oxidase enzyme activity through the modulation of channel pathways or through different enzymatic systems, in order to reduce the production of reactive oxygen species (ROS). This antioxidant potential is in turn associated with the location and amount of OH groups contained in the structure that modifies the thermodynamics, favoring the antioxidant site [30].

Moreover, tannins feature antioxidant activity due to diferent polyphenols types; for example, the most active are dimeric procyanidins that scavenge ABTS+ radical, hypochlorous acid, along the reducing power evaluation [23]. Nevertheless, tannins can affect the inflammatory response via their radical scavenging activities such as the production of superoxide, hydroxyl radical, and other oxidizing species that initiates a feedback cycle. The radical products of inflammatory reactions serve as inducing agents for further inflammation [30].

Studies performed on binding flavonoids modes with COX-2 explored that some flavonols and flavones containing a 2, 3-double bond may act as preferential inhibitors of COX-2. Good anti-inflammatory activity and reduced gastrointestinal side effects were shown [29].

## 7. Polysulfides and thiosulfinates

Plants of the genus *Allium* like garlic, onion, and leek have been used since ancient times for their numerous health benefits [32, 33]. The multiple therapeutic properties are due to the sulfur compounds which are obtained primarily from garlic such as alliin, allicin, ajoene, allylpropyl disulfide, diallyl trisulfide, S-allylcysteine, and S-allylmercaptocysteine [34].

Unlike other plants, those belonging to the genus *Allium* do not have sulfur compounds available for extraction, purification, and subsequent use. Here, the synthesis must be induced as they are formed in an enzymatic reaction from alliin, who subsequently meets the enzyme alliinase that is released from the vacuoles of plant cells after some type of mechanical stress (cut, grind, or crush). Once the enzymatic reaction has been carried out, thiosulfinates (diallyl sulfide, diallyl disulfide, and diallyl tetrasulfide) and polysulfides (diallyl tetrasulfide, S-allylmercaptocysteine, S-allylmercaptoglutathione, allyl mercaptan, allyl persulfide, and dithiins) are obtained as products. The structure is shown in **Figure 9**. Therapeutic properties such as antiviral, antimicrobial, antifungal, anticancer, antihypertensive, antioxidant, and anti-inflammatory are attributed to them [35–37].

The great antioxidant plants capacity of the genus *Allium* is given by the synergistic relationship that exists between the sulfur compounds and polyphenols present in these species [38]. The interaction of those compounds with uric acid acts directly on the molybdopterin (Mo-Pt) domain that causes the antioxidant action [39]. Recent studies propose different models of antioxidant action mechanisms such as the decrease in the production of free radicals or models where it is presented that by increasing the levels of glutathione (GSH) in the cell. The oxidized glutathione (GSSH) levels are decreased in the cell due to the increase in the enzymatic activity of GSSG reductase or superoxide dismutase. Those effects have been attributed to garlic [40]. Reduced glutathione (GSH) serves as a reducer for glutathione peroxidase (GPX), a natural protective enzyme against free radicals in the company of superoxide dismutase and catalase [34]. **Table 1** shows the thiosulfinates and polysulfides reported with antioxidant and anti-inflammatory activity, the respective action mechanism.

As anti-inflammatories, on the contrary, sulfur compounds act by suppressing the expression of inducible nitric oxide (NO) synthase and sustaining the NO produced by endothelial NO synthase that has demonstrated effectiveness against inducible diseases and inflammatory disorders. On the other hand, the anti-inflammatory garlic effects were linked to a direct decrease in the activation of the toll-like receptor

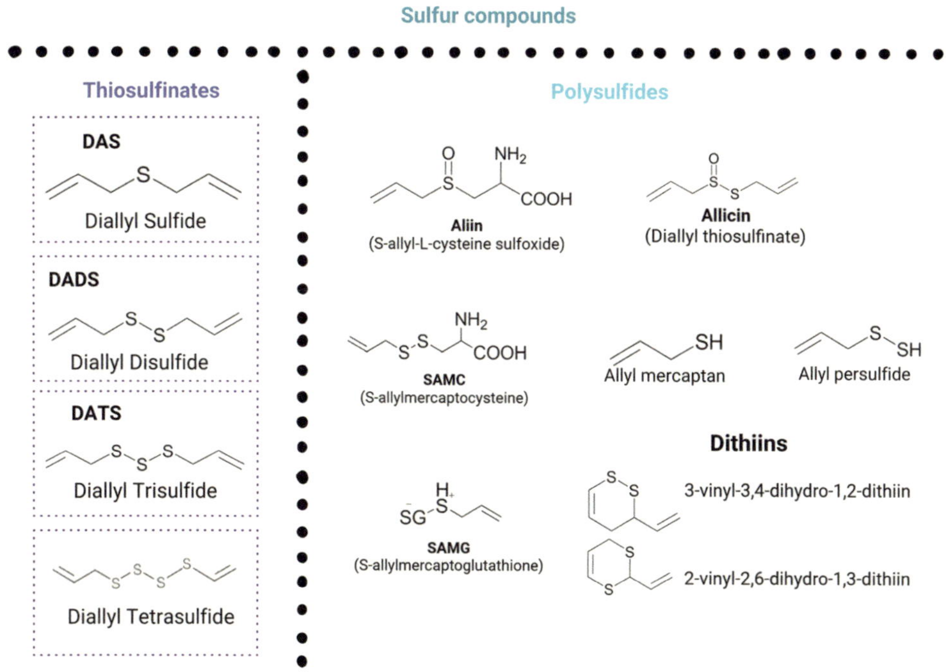


**Figure 9.**
*Thiosulfinates (diallyl sulfide, diallyl disulfide, diallyl trisulfide, diallyl tetrasulfide, etc.) and polysulfides (aliin, allicin, s-allylmercaptoglutathione, a-allylmercaptocysteine, allyl mercaptan, allyl persulfide, dithiins, etc.) chemical structure. Source: PubChem, 2024. Own modification [7].*

| Compound | Pharmacological activities | Mechanism of action |
|---|---|---|
| Alliin (S-allyl cysteine sulfoxide) | Antioxidant<br>Anti-inflammatory | Suppresses the formation of ROS and down-regulates the expression of NADPH oxidase 1 (Nox1) |
| Diallyl sulphide (DAS) | Antioxidant<br>Anti-inflammatory | Decreases the production of ROS Increases cleaved caspase 3 and p38 expression |
| Diallyl disulfide (DADS) | Anti-inflammatory | Modulates immune cells and NF-kB signaling pathway |
| Diallyl trisulfide (DATS) | Antioxidant<br>Anti-inflammatory | Increases glutathione and antioxidant enzymes and modulates ROS. Suppressed the synthesis of pro-inflammatory nitric oxide. |
| Allyl methyl trisulfide | Antioxidant | Decreases oxidative stress through the enhancement of the antioxidant system |
| Thiacremonone | Anti-inflammatory | It blocks the NF-kB activity |
| Ajoenes | Antioxidant | Superoxide inhibition |
| S-allyl cysteine (SAC) | Antioxidant | Scavenges reactive oxygen species (ROS), induces antioxidant enzymes, and inhibits peroxidant enzymes like xanthine oxidase and cyclooxygenase. |

**Table 1.**
*Action mechanism of thiosulfinates (diallyl sulfide, diallyl disulfide, diallyl trisulfide, diallyl tetrasulfide, etc.) and polysulfides (aliin, allicin, s-allylmercaptoglutathione, a-allylmercaptocysteine, allyl mercaptan, allyl persulfide, dithiins, etc.) reported with antioxidant and anti-inflammatory activity. Source: Jikah & Okoro, 2023. Own modification [41, 42].*

4 (TLR4) signaling cascade in macrophages and modulation of immune cells, by the reduction in nuclear factor Kappa B (NF -kB) levels and enzymes like glycogen synthase kinase (GSK)-3, the expression of pro-inflammatory cytokines IFN-y and TNF-a decreased after a single application of garlic essential oil, thereby preventing prolonged inflammation, cell damage, and tissue injury [39, 41–43].

## 8. Conclusions

Inflammation is a body's natural response to infectious agents and injuries. However, this can be harmful when it becomes chronic causing cellular damage and contributing to the development of several diseases. Along this chapter, it was discussed that the metabolism of plants is complex and follows a perfect synchrony in conjunction with the other metabolic processes of the plant, in order to provide it with the protection and adaptability to its environment, which it requires, thanks to phytochemicals derived from the secondary metabolism of plants. These compounds offer numerous health benefits to humans by making them versatile options that help to complement conventional medicine. Given their availability, accessibility, their disease prevention potential, along with their sustainability and lower environmental impact compared to synthetic medicines, makes these phytochemicals a considerable option for diseases treatment.

Phytochemicals, such as phenols, polyphenols, tannins, flavonoids, and more recently thiosulfinates and polysulfides, play an important role as anti-inflammatories and antioxidants in humans. Neutralizing free radicals cause oxidative stress, contribute to the inflammatory process, and act directly in modulating the activity of various enzymes and inflammatory pathways in the body. Helping to reduce inflammation and mitigate its own symptoms, such as pain and swelling. Making the use of these ancient natural compounds in conjunction with new technologies and scientific advances to contribute on technological development of new medications and devices for the treatment of vulnerable populations.

## Acknowledgements

Thanks to the National Research Council of Humanities, Sciences and Technologies (CONAHCYT) for financing my doctoral studies through a scholarship with the number 1309991, to the Metabolic Engineering and Bioinformatics Laboratory of the University of Guadalajara, to my supervisors for their support and guidance, and to my student for her wonderful work and effort.

## Conflict of interest

The authors declare that they have no known competing financial interests or personal relationships that could have influenced the work presented in this book chapter.

## Acronyms and abbreviations

| | |
|---|---|
| NF-kB | nuclear factor kappa B |
| TNF-a | tumor necrosis factor alpha |

| | |
|---|---|
| IL-1 | interleukin-1 |
| IL-6 | interleukin-6 |
| IL-8 | interleukin-8 |
| 5-HTP | 5-hydroxytryptophan |
| A2 | phospholipases |
| COX | cyclooxygenases |
| PGs | prostaglandins |
| LOX | lipoxygenases |
| MVA | mevalonate |
| HMGR | 3-hydroxyl-3-methylglutaryl co-enzyme A reductase |
| MVK | mevalonate kinase |
| HIDS | hyperimmunoglobulin D periodic fever syndrome |
| MA | mevalonate aciduria |
| ROS | reactive oxygen species |
| iNOS | inducible nitric oxide synthase |
| PPARy | peroxisome proliferator-activated receptor gamma |
| PI 3-kinase | phosphoinositide 3-kinase |
| AA | arachidonic acid |
| HpETEs | hydroperoxyeicosatetraenoic acids |
| HETEs | hydroxyeicosatetraenoic acids |
| HT | hydrolysable tannins |
| S-allyl cysteine sulfoxide | alliin |
| Nox1 | NADPH oxidase 1 |
| DAS | diallyl sulfide |
| DADS | diallyl disulfide |
| DATS | diallyl trisulfide |
| SAC | S-allyl cysteine |
| Mo-Pt | molybdopterin |
| GSH | glutathione |
| GSSH | oxidized form of glutathione |
| GPX | glutathione peroxidase |
| NO | nitric oxide |
| TLR4 | toll-like receptor 4 |
| (GSK)-3 | glycogen synthase kinase |

## Author details

Ana Montserrat Corona-España[1*†], Mario Alberto Garcia-Ramirez[2†], Regina Romo-Gonzalez[3†], Ingrid Mayanin Rodriguez-Buenfil[4†] and Orfil González Reynoso[5†]

1 Department of Chemistry, University Center of Applied Sciences and Engineering, University of Guadalajara, Guadalajara, Mexico

2 Department of Electro-Photonic Engineering, University Center of Applied Sciences and Engineering, University of Guadalajara, Guadalajara, Mexico

3 University Center of Applied Sciences and Engineering, University of Guadalajara, Guadalajara, Mexico

4 Center for Research and Assistance in Technology and Design of the State of Jalisco. A.C Southeast Sub-Headquarters, Merida, Mexico

5 Metabolic Engineering and Bioinformatics Laboratory, Department of Chemical Engineering, University Center of Applied Sciences and Engineering, University of Guadalajara, Guadalajara, Mexico

*Address all correspondence to: ana.corona@academicos.udg.mx

† These authors contributed equally.

IntechOpen

© 2024 The Author(s). Licensee IntechOpen. This chapter is distributed under the terms of the Creative Commons Attribution License (http://creativecommons.org/licenses/by/4.0), which permits unrestricted use, distribution, and reproduction in any medium, provided the original work is properly cited.

## References

[1] Apostolopoulos V, Vivek JF, Chavda P. Inflammation: The Cause of all Diseases. Vol. 9. Switzerland: MDPI; 2023. ISBN 978-3-0365-8823-0. DOI: 10.3390/books978-3-0365-8823-0

[2] Lic Maureen Hernández Ángel Luis García Bacallao Lourdes Rojo Domínguez Delia Mercedes, García Gómez Vicente. Plantas con propiedades antiinflamatorias. 2022. Available from: http://personal.redestb.es/martin/pfito.htm

[3] Gonzáles G, Caballero-Gutiérrez L. Acta médica Peruana Amp Alimentos Con Efecto Anti-Inflamatorio Foods with Anti-Inflammatory Effect. Perú: Colegio Médico del Perú; 2016

[4] Stone WL, Basit H, Burns B. Pathology, Inflammation. Treasure Island (FL): StatPearls Publishing; 2024

[5] Soon G. General Pathology Inflammation and Repair. Singapore: National University Hospital Main Building; 2002

[6] Filella X, Molina R, Ballesta AM. Funciones de las citocinas. 2002. Available from: https://acortar.link/HRDj6q

[7] Pubchem. Molecules. 2024. Available from: https://pubchem.ncbi.nlm.nih.gov/

[8] Abdulkhaleq LA, Assi MA, Abdullah R, Zamri-Saad M, Taufiq-Yap YH, Hezmee MNM. The crucial roles of inflammatory mediators in inflammation: A review. Veterinary World. 2018;**11**(5):627-635. DOI: 10.14202/vetworld.2018.627-635. ISSN 09728988

[9] Manisha WH, Rajak R, Jat D. Oxidative stress and antioxidants: An overview. International Journal of Advanced Research and Review. 2017;**2**:110-119

[10] John Betteridge D. What is oxidative stress? Metabolism. 2000;**49**(2):3-8. DOI: 10.1016/S0026-0495(00)80077-3. ISSN 00260495

[11] Venereo JR, Gutiérrez. DaÑo oxidativo, radicales libres y antioxidantes. Revista Cubana de Medicina Militar. 2002;**31**:33-126. Available from: http://scielo.sld.cu/pdf/mil/v31n2/mil09202.pdf

[12] Sies H. Oxidative stress: Concept and some practical aspects. Antioxidants. 2020;**9**(9):852. DOI: 10.3390/antiox9090852. ISSN 2076-3921

[13] Mendoza N, Escamilla EM, Silva. Introduction to Phytochemicals: Secondary Metabolites from Plants with Active Principles for Pharmacological Importance. Vol. 11. London, UK: InTech; 2018. DOI: 10.5772/intechopen.78226

[14] Carril AÁGEP-U. Metabolismo secundario de plantas. Reduca (Biología). Serie Fisiología Vegetal. 2009;**2**:119-145. ISSN 1989-3620

[15] Copeland L, Turner JF. The Regulation of Glycolysis and the Pentose Phosphate Pathway. Cambridge, Massachusetts, United States: Academic Press, Elsevier; 1987. pp. 107-128. DOI: 10.1016/B978-0-12-675411-7.50010-0

[16] Kougan GB, Tabopda T, Kuete V, Verpoorte R. Simple Phenols, Phenolic Acids, and Related Esters from the Medicinal Plants of Africa. Cambridge, Massachusetts, United States: Elsevier; 2013. pp. 225-249. DOI: 10.1016/B978-0-12-405927-6.00006-0

[17] Nissen S, Namuth D, Hernández-Rios I. Inhibidores de la síntesis de aminoácidos aromáticos. Plant and Soil Sciences ELibrary; 2018. Available from: https://passel2.unl.edu/view/lesson/9a49f4ad2e86

[18] Karlic H, Varga F. Mevalonate Pathway. Cambridge, Massachusetts, United States: Elsevier; 2017. DOI: 10.1016/B978-0-12-801238-3.65000-6

[19] Dewick PM. Medicinal Natural Products a Biosynthetic Approach. 3rd ed. Hoboken, Nueva Jersey: John Wiley & Sons; 2009. ISBN 978-0-470-74168-9. Available from: https://ipa-pasca.unpak.ac.id/pdf/Dewick_Medicinal_natural_products.pdf

[20] Cho S-H, Tóth K, Kim D, Vo PH, Lin C-H, Handakumbura PP, et al. Activation of the plant mevalonate pathway by extracellular ATP. Nature Communications. 2022. ISSN 2041-1723.;**13**(1):450. DOI: 10.1038/s41467-022-28150-w

[21] Mazimba O, Nana F, Kuete V, Singh GS. Xanthones and Anthranoids from the Medicinal Plants of Africa. Cambridge, Massachusetts, United States: Elsevier; 2013. pp. 393-434. DOI: 10.1016/B978-0-12-405927-6.00011-4

[22] Belščak-Cvitanović A, Durgo K, Huđek A, Bačun-Družina V, Komes D. Overview of Polyphenols and their Properties. Cambridge, Massachusetts, United States: Elsevier; 2018. pp. 3-44. DOI: 10.1016/B978-0-12-813572-3.00001-4

[23] Macáková K, Kolečkář V, Cahlíková L, Chlebek J, Hoštálková A, Kuča K, et al. Tannins and their Influence on Health. Cambridge, Massachusetts, United States: Elsevier; 2014. pp. 159-208. DOI: 10.1016/B978-0-12-803961-8.50006-3

[24] Alonso MGD. Los compuestos fenólicos, un acercamiento a su biosíntesis, síntesis y actividad biológica. Revista de Investigación Agraria y Ambiental. 2018. ISSN 2145-6453;**9**(2):81-104. DOI: 10.22490/21456453.1968

[25] Zhang H, Tsao R. Dietary polyphenols, oxidative stress and antioxidant and anti-inflammatory effects. Current Opinion in Food Science. 2016;**8**(4):33-42. DOI: 10.1016/j.cofs.2016.02.002. ISSN 22147993

[26] Carmela S, Roberta M, Carmela F, Roberta DB, Beatrice S, Rosaria V. Polyphenols, intracellular signalling and inflammation. Annali dell Istituto Superiore di Sanità. 2007;**43**(2):394-405

[27] Sieniawska E, Baj T. Tannins. Cambridge, Massachusetts, United States: Elsevier; 2017. pp. 199-232. DOI: 10.1016/B978-0-12-802104-0.00010-X

[28] Diane JM. Tannins as Anti-Inflammtory Agents. Miami University; 2006. pp. 1-12. Available from: https://etd.ohiolink.edu/acprod/odb_etd/ws/send_file/send?accession=miami1154451707&disposition=inline

[29] Panche AN, Diwan AD, Chandra SR. Flavonoids: An overview. Journal of Nutritional Science. 2016. ISSN 2048-6790;**5**(12):e47. DOI: 10.1017/jns.2016.41

[30] Ullah A, Munir S, Badshah SL, Khan N, Ghani L, Poulson BG, et al. Important flavonoids and their role as a therapeutic agent. Molecules. 2020;**25**(11):5243. DOI: 10.3390/molecules25225243. ISSN 1420-3049

[31] Dildo ML, Álvaro SL. El uso de taninos condensados como alternativa nutricional y sanitaria en rumiantes. Revista de Medicina Veterinaria. 2008;**16**(12):87

[32] Guillamon E. Efecto de compuestos fitoquímicos del género allium sobre el sistema inmune y la respuesta inflamatoria. Ars Pharmaceutica (Internet). 2018. ISSN 0004-2927;**59**:9. DOI: 10.30827/ars.v59i3.7479

[33] Hall A, Troupin A, Londono-Renteria B, Colpitts TM. Garlic organosulfur compounds reduce inflammation and oxidative stress during dengue virus infection. Viruses. 2017;**9**:7. DOI: 10.3390/v9070159. ISSN 19994915

[34] López VR. Allium sativum como fuente potencial de molÉculas anticancerÍgenas [Thesis]. Universidad Complutense; 2016. pp. 2-16

[35] Mehra R, Jasrotia RS, Mahajan A, Sharma D, Iquebal MA, Kaul S, et al. Transcriptome analysis of snow mountain garlic for unraveling the organosulfur metabolic pathway. Genomics. 2020;**112**(1):99-107. DOI: 10.1016/j.ygeno.2019.07.014. ISSN 10898646

[36] Poojary MM, Putnik P, Kovačević DB, Barba FJ, Lorenzo JM, Dias DA, et al. Stability and extraction of bioactive sulfur compounds from allium genus processed by traditional and innovative technologies. Journal of Food Composition and Analysis. 2017;**8**:28-39. ISSN 08891575

[37] Stoll A, Seebeck E. Chemical Investigations on Alliin, the Specific Principle of Garlic. Vol. 11. Hoboken, Nueva Jersey: Interscience Publishers Inc.; 1951. pp. 377-400. DOI: 10.1002/9780470122563.ch8

[38] Tran A-SN, Pham VH, Duong CD, Bui QTP. Extraction conditions, chemical composition and biological activity of essential oil of allium schoenoprasum l. bulb from Quang Tri province, Vietnam. Food Chemistry Advances. 2024;**4**(6):100574. DOI: 10.1016/j.focha.2023.100574. ISSN 2772753X

[39] Kurnia D, Ajiati D, Heliawati L, Sumiarsa D. Potential of flavonol and sulfur compounds from allium leaves as an antioxidant and xanthine oxidase inhibition in silico study. Food Chemistry Advances. 2023;**3**:12. DOI: 10.1016/j.focha.2023.100383. ISSN 2772753X

[40] Heidi Ramírez-Concepción R, Castro-Velasco LN, Martínez-Santiago E. Artículos de revisión efectos terapéuticos del ajo (allium sativum). Salud y Administración. 2016;**3**(8):39-47

[41] Jikah AN, Edo GI. Mechanisms of action by Sulphur compounds in allium sativum. A review. Pharmacological Research - Modern Chinese Medicine. 2023;**12**:100323. ISSN 26671425

[42] Okoro BC, Dokunmu TM, Okafor E, Sokoya IA, Israel EN, Olusegun DO, et al. The ethnobotanical, bioactive compounds, pharmacological activities and toxicological evaluation of garlic (allium sativum): A review. Pharmacological Research - Modern Chinese Medicine. 2023;**9**:100273. ISSN 26671425

[43] Jafari F, Khalilzadeh S, Nejatbakhsh F, Naderi M. Therapeutic effects of garlic (allium sativum) on female reproductive system: A systematic review. Heliyon. 2023;**9**:12. DOI: 10.1016/j.heliyon.2023.e22555. ISSN 24058440

Section 3

# Bioactive Potential of Phytochemicals

Chapter 6

# Bioactive Compounds and Biological Activities of *Curculigo orchioides* Gaertn

*Nguyen Van Khiem, Nguyen Minh Duc and Dinh Thanh Giang*

## Abstract

*Curculigo orchioides* Gaertn. is a rare herb belonging to the family Hypoxidaceae, popularly known as "Kali Musli" in India, "Sam Cau", "Tien Mao" in Vietnam, "Xian Mao" in China. This species is native and widely distributed in Asian countries such as India, Nepal, Pakistan, South China, Laos, Vietnam, Myanmar, and the Philippines. *C. orchioides* contains several classes of chemicals, including saponins, alkaloids, flavonoids, phytosterols, steroids, free reducing sugars starch, proteins, resins, tannins, mucilage, fats, and inorganic compounds. Rhizoma Curculiginis is used in traditional medicine remedies in various countries for the treatment of a range of ailments, including impotency, aphrodisiacs, tonics, jaundice, and skin conditions. The scientific evidence indicates its aphrodisiac, spermatogenic, estrogenic, immunomodulatory, antioxidant, anti-inflammatory, antiosteoporosis, antidepressant, hepatoprotective, nephroprotective, antidiabetic, antimicrobial. Antiasthmatic, antistress, anticancer, antigout, antihypertensive, antimalarial, and cardiovascular activities. As traditional herbal medicine, *C. orchioides* has been investigated for its pharmacological activity against diabetes, fungi, bacteria, inflammation, osteoporosis, oxidative stress, cancer, and neurodegeneration. Findings of bioactive compounds and the biological effects of the extract of *C. orchioides* provide scientific support for the folk use of this plant in the traditional, unvalidated herbal drugs as aphrodisiac and rejuvenating tonics.

**Keywords:** *Curculigo orchioides*, bioactive compounds, biological activity, sam cau, curculigoside, orcinol glucoside

## 1. Introduction

Globally, the use of products derived from medicinal plants is increasing for the prevention and treatment of various human diseases [1]. According to a survey conducted in the UK and Saudi Arabia, about 80% of the population uses medicinal plants for primary health care and to treat diseases in humans [2, 3]. In 2021, herbal medicine consumption in the world market reached a value of 230.03 billion USD, with a predicted rise to 430.05 billion USD in 2028 [4]. Factors, including availability, low-cost, fewer side effects, and the efficacy of herbal medicines are the main reasons for humans have surged using them worldwide [5]. Generally, herbal medicine is safe for the treatment of various diseases in humans. In fact, medicinal plants contain different compound groups, including saponin, phenolic, flavonoid, alkaloid, and terpenoid. They act individually or in combination with other compounds, giving beneficial or harmful health effects for humans.

*C. orchioides* is a well-known medicinal plant in Asia. It has commonly called "Kali Musli" in India and "Xian Mao" in China [6], "Sam Cau" or "Tien Mao" in Vietnam [7]. It is one of the common herbs, used in traditional medicines in Asian countries [8]. The rhizome of *C. orchioides* is used widely in traditional medicines. In India, *C. orchioides* is used for the treatment of strengthening the spleen, aphrodisiac, antiaging, weight loss, menstrual irregularities, menstrual cramps and amenorrhea, kidney, and bones [9]. Wang et al. [10] reported the main phytoconstituents of *C. orchioides* that were isolated from the rhizome including glycosides, alkaloids, saponins, and polysaccharides. Curculigoside, a phenolic glycoside was the main component, having biological activities such as antiosteoporotic and neuroprotective [11, 12]. Different extracts of *C. orchioides* were proven anti-inflammatory, antidiabetic, aphrodisiac, antioxidant, and antimicrobial activities [13–16], and as a tonic for overcoming impotence [17].

Currently, *C. orchioides* is provided a limited amounts, due to the unsustainable harvesting of genetic resources, and shrinking of its' natural habitats. It is ranked at high risk of extinction soon in Vietnam [18]. *C. orchioides* is a near-threatened medicinal herb as per the International Union for Conservation of Nature (IUCN) status (2012) [19]. In recent years, there has been research on cultivation providing medicinal materials for medicinal products derived from *C. orchioides* [20–23]. Bioactive compounds and biological activity findings will provide scientific support the folk use of this plant to the traditional unvalidated herbal drugs as an aphrodisiac and rejuvenating tonics. In this chapter, we describe the phytochemicals of *C. orchioides* and its biological activities.

## 2. Botanical details

*C. orchioides* belongs to the family Hypoxidaceae. It is known as Golden Eye Grass or Black Musale in English, Kalimusli in Hindi. The genus *Curculigo* consists of 10 species, distributed in India, Sri Lanka, Japan, Malaysia, Vietnam, Australia, etc. [7, 19]. *C. orchioides* originated in Asia. It is distributed in plains and shady forests. It grows well on moist fertile and laterite soils, in rock crevices from sea level to 2300 m altitude. It occurs in the subtropical and tropical regions. The raw materials are supplied for the drug industry, collected mainly from wild genetic sources, and it is depleting the natural populations of these medicinal plants [24].

*C. orchioides* is found almost all over midland and mountainous regions covering the North, Central, and South provinces of Vietnam. It can grow well in shade (25%) and the areas receiving high rainfall [7]. *C. orchioides* is a small herbaceous species. It has an elongated rhizome (5–20 cm long) with a fleshy lateral root system. The rhizome is often vertical (**Figure 1A**). Leaves are very variable, from narrowly linear to lanceolate, acute, plicate, or flat. They are crowded on the rhizome-bearing sheathing leaf bases. The petiole is green or purple-red, short, or absent. Flowers appear throughout the year, but mostly from May to July, light yellow, bisexual, sessile, regular (**Figure 1B**). Perianth is six-lobed, lobes yellow, stamens-6, filaments (~ 2 mm

**Figure 1.**
*Morphology of* C. orchioides *plant: (A) Rhizome-bearing leaflets, flower, fruit, and lateral roots; (B): flower; (C): Seed-bearing ripened fruits; (D): Seeds.*

long), filiform, anthers are approximately 2 mm, ovary 3-celled, oblong. Ovules are numerous per cell, stigma-3, lobes elongate. Fruits are oblong (**Figure 1C**). The seeds are black, and have different shapes such as spheres, beaks, and deep grooves in wavy lines (**Figure 1D**) [24, 25].

## 3. Chemical compounds

Several researchers found that the rhizomes of *C. orchioides* contain several classes of chemicals such as alkaloids, flavonoids, phytosterols, free reducing sugars, glucuronic acid, glycoside (including cardiac glycosides, saponins, steroids, polysaccharides (as hemicellulose and other polysaccharides), starch, proteins, resins, tannins, mucilage, fats, calcium oxalate, inorganic compounds, and fibers [26, 27].

Mehta et al. [28] demonstrated that the hexane extract of *C. orchioides* containing alkaloid-lycorine, sterols (β-sitosterols), and sapogenin (yuccagenin), the flavone (5,7 dimethoxy glucopyranoside), fatty acids (linolenic, palmitic, behenic, arachidic, and oleic). 5-hydroxy-2-0-b-d-glucopyranosyl benzyl 1,2,6-dimethoxy benzoate, a curculigoside was isolated from the rhizomes [29]. Other studies showed that *C. orchioides* rhizome also contain a cycloartane, namely, curculigol [30], corchiocide

A (as 25-hydroxy-33- methylpentatriconta-6-one, 21-hydroxytetracontan-20-one, 27-hydroxy-tricontan-6-one, 2–methoxy-4-acetyl-5-methytricontane) [31], curculigol, curculigenin A, Curculigo saponin AF, cycloartenol, sitosterol, and stigmasterol [32]. Several researchers isolated aliphatic hydroxy ketones, 27-hydroxy tricontane-6-one, 23-hydroxy tricontane-2-one, 21-hydroxy tetracontan-20-one and 4-methyl heptadecanoic acid from the rhizome of *C. orchioides* [33, 34].

It is found that the rhizome of *C. orchioides* has a phenolic glycoside, namely orcinol-3-bD-xylopyranosyl-(1®6)- β-D-glucopyranoside and Hentriacontanol [35–37], aliphatic compounds such as 25-dihydroxy-33-methyl pentatricontanone and natural triterpene curculigol as 24-methyl cycloart-7-en-3-beta-20-diol [38]. Next, Xu et al. [39] isolated from the rhizomes of *C. orchioides* a pentacyclic triterpene, characterized as 31-methyl-3-oxo-20-ursen-28-oic acid. Following, Curculigo saponin A-M, glycoside 5, 7-dimethoxy myricetin 3-0-L-xylo- pyranosyl 4-0-b-D glycopyranoside, free sugars, mucilage, hemicellulose, and other polysaccharides are found be present in rhizomes of *C. orchioides* [40]. There are about 111 secondary metabolites, three proteins, and two polysaccharides, identified from *Curculigo* plants [41]. Compared to *Curculigo* species, their extracts contain various compounds, that includes phenolic glycosides, phenols, lignan glycosides, lignans, triterpenoid glycosides, triterpenes, polysaccharides, flavones, aliphatic compounds, alkaloids, eudesmanes, and bioactive proteins as neoculin, curculin, and β-amylase [9]. Besides, alkaloids, phenols, flavonoids, and tannins were present in the methanol, ethanol, and chloroform extracts of *C. orchioides,* but they are absent in the ethyl acetate extract [42].

It is demonstrated that methanolic extracts of *C. orchioides* rhizome contain alkaloids, carbohydrates, steroids, saponins, phenols, tannins, and flavonoids [13, 26]. While O-acetyl-glucomannan COP90-1, a compound soluble in water is isolated from *C. orchioides* [43] and COP70-3, a homogeneous heteropolysaccharide compound was found in the rhizomes of *C. orchioides* [44]. Other findings showed that phenolic glucosides were identified in *C. orchioides* rhizomes, including curculigoside H and orcinoside I [45], curculigoside A, B, C, G, orcinol glucoside B, benzyl-O-β-D-glucopyranoside, 3-hydroxy-5-methylphenol-1-O-[β-glucopyranosyl-(1-6)-β-Dglucopyranoside], glucosyringic acid, 3-hydrox-5-methyl-phenol-1-O-[β-apiosyl-(1-6)-β-glucopyranoside], and orcinol glucoside [46], 3-(4-hydroxy-3-methoxyphenyl) propane-1,2-diol, 3-(4-hydroxy-3,5-dimeth-oxyphenyl) propane-1,2-diol, piperoside, 4-ally-2, and 6-dimeoxy phenol glucoside [47]. Niu et al. [48] isolated 3,5-Dihydroxy-4-methoxybenzoic acid and p-Hydroxycinnamic acid from powdered rhizomes. *C. orchioides* contains cuculligines, chlorophenolic glucosides, B, C, and curculigine D [40, 49, 50]. *C. orchioides* rhizome obtains 11 chlorophenolic glucosides, such as curculigine E–G, I, and K–O [45, 46, 51], chlorophenolic glucosides named ascurculigine P and Q [52]. *C. orchioides* contains also cycloartane-type triterpenoid ketone [53], terpenoids including 3S,5R,6S,7E,9R)-megatigma-7-ene-3,5,6,9-tetrol, actinidioionoside, (6S,9R)-roseoside, (−)- angelicoidenol-2-O-β-D-glucopyranoside, (−)-angelicoidenol-2-O-β-apiofuranosyl-(1→6)-β-D-glucopyranosidetetillapyrone [(7R,9S,10R)-3-methyl-5-(4-hydroxyl-5-hydroxylmethyltetrahydrofuryl)]-6-hydroxypyran-2-one [54, 55].

*C. orchioides* rhizome contains also cyclodipeptides, including cyclo-(L-Ala-L-Tyr), cyclo- (Gly-D-Val), cyclo-(LeuAla), cyclo-(Val-Ala), cyclo-(L-Ser-L-Phe), cyclo-(LeuThr), cyclo-(S-Pro-RLeu), and cyclo- (Leu-Ser) [47]. Several studies found that the aqueous leaf extract of *C. orchioides* contains different alkaloids and phenolic compounds [56]. Leaf methanolic extract of *C. orchioides* contains the compounds such as 2-Myristynoyl pantetheine, 9-Hexadecenoic acid, 9,10-Secocholesta-5,7,10(19)-triene-3,24,25-triol, (3á,5Z,7E)-, Cyclobarbital, Thieno[2,3-c] furan-3-carbonitrile,2-amino-4,6-dihydro-4,4,6,6-tetramethyl-, Testosterone, 17-O-(t-butyldimethylsilyl), 2-Myristynoyl-glycinamide, N-[3,5-Dinitropyridin-2-yl] glutamic acid, 4-Dehydroxy-N-(4,5-methylenedioxy-2-nitrobenzylidene)tyramine, 4-Acetyloxyimino-6,6-dimethyl-3-methylsulfanyl-4,5,6,7-tetrahydrobenzo[c] thiophene-1-carboxylic acid methyl ester. Meanwhile, rhizome methanolic extract contains as 2,7-Diphenyl-1,6-dioxopyridazino [4,5:2′,3′]pyrrolo[4′,5′-d]pyridazine, sarreroside, 9-Octadecenoic acid, (2-phenyl-1,3-dioxolan-4-yl)methyl ester, cis-, Ergosta-5, 22-dien-3-ol, acetate, (3á,22E)-, Digitoxin, Ethyl iso-allocholate, Rhodopin, 10-Secocholesta-5,7,10(19)-triene-3,24,25-triol, Cyclobarbital [57]. In a study, three phenolic glycosides from the ethanol extracts of the rhizomes of *C. orchioides*, that collected in Central Vietnam. They were structurally identified as orcinol glucoside, orcinol-1-O-(6’-O-acetyl)-α-D-glucopyranoside, and curculigoside [58]. *C. orchioides* harvests in Thua Thien Hue province, Central Vietnam found to have high orcinol glucoside content, used as standard compounds for quality of medicinal herbs [59]. Yadav et al. [60] demonstrated that *C. orchioides* alcoholic extract containing compounds including d-Lyxo-d-manno-nononic-1,4-lactone, 3′,8,8′-Trimethoxy-3-piperidyl-2,2′-binaphthalene-1,1′,4,4′-tetrone, 7-Methyl-Ztetradecen-1-ol acetate, Paromomycin, Geranyl isovalerate, Tert-Hexadecanethiol, 1,2-Propanediol, 3-(tetradecyloxy), n-Hexadecanoic and 9-Octadecenoic acids, (2-phenyl-1,3-dioxolan-4-yl) methyl ester, trans, 2H-Cyclohepta[b] furan-2-one, 6-[1-(acetyloxy)-3-oxobutyl] -3,3a,4,7,8,8a-hexahydro-7-methyl-3-methylene. In which, several compounds were activity assessed.

## 4. Biological activities

### 4.1 Aphrodisiac activity

Sexual behaviors of male rats such as penile erection, mating performance, mount frequency, and mount latency are significantly improved by using a dose of 100 mg/kg body weight of ethanolic rhizome extract of *C. orchioides*. Besides, the lyophilized aqueous extract of the rhizome may improve significantly sexual activities of male rats at a dose of 200 mg/kg [61].

### 4.2 Spermatogenic activity

Orientation behaviors and spermatogenesis of male rats are significantly increased at a dose of 100 mg/kg body weight of ethanolic extract of *C. orchioides* rhizomes. Ethanolic extract effects on orientation behaviors of males toward females, environment, self, and type of mobility. It effects also on spermatogenesis, increasing in number of spermatocytes and spermatids in albino rats [62]. Thanh-Nhan Bui-Le et al. [63] found the *C. orchioides* collected in Vietnam reduced the heat stress-induced spermatogenesis complications in mice. Extract of rhizome increased spermatogenic cells and spermatozoa in heat-exposed mice compared to heat-exposed mice without eating extract of *C. orchioides*. Otherwise, the aqueous extract of rhizome may enhance the serum total testosterone content in heat-exposed mice. Thus, *C. orchioides* was effective against heat stress-induced spermatogenesis complications in the murine model.

### 4.3 Estrogenic activity

*C. orchioides* expresses estrogenic activity. Thakur et al. [64] found that *C. orchioides* reduce hyperglycemia levels, induced by streptozotocin and improves sexual dysfunction in male rats. Sexual behaviors in males, such as penile erection index, seminal fructose, and sperm content are recorded. Steroid synthesis and restored sexual function are enhanced by the spermatogenesis process. *C. orchioides* may also improve better hormone absorption into the gonads [61]. Total glucosides of *C. orchioides* increase the thymus, uterus, and spleen indices, testosterone, and estrogen contents. It may decrease luteinizing hormone contents in perimenopause model mice [65, 66].

### 4.4 Immunomodulatory activity

*C. orchioides* ethanolic extract improved adaptation to hypoxia and high temperatures. Extracts from *C. orchioides* were found to be sedative, anticonvulsant, and androgen-like. Immunological activity increased in mice, as well [67]. Lakshmi et al. [68] isolated, and purified two *C. orchioides* phenolic glucosides and found significant immuno-stimulant activity. They increase hemagglutination titers and delay hypersensitivity responses. Zhou et al. [69] showed polysaccharides of *C. orchioides*, that stimulate splenocyte proliferation, but they did not affect thymocyte proliferation. Its polysaccharides inhibit thymocyte and spleenocyte proliferation in the experimental model using ConA-induced splenocyte proliferation. It suggests that polysaccharides in *C. orchoides* are the key immune-modulating components. Another study found that the methanolic extract of *C. orchioides* may prevent the cytotoxic effects of drugs. Humoral antibodies and leukocyte levels increased, depending upon the methanol extract dose, when it was applied to cyclophosphamide-induced immunosuppressed or normal mice [70]. Triterpene saponin compound, Acycloartane-type of *C. orchioides* rhizome was also found to enhance the number of lymphocytes of the spleen remarkably in mice without affecting antibody induction [71]. Mice's immune function was enhanced by polysaccharides. Polysaccharides increase spleen, thymus, and hemolytic indexes in the serum of normal mice [72].

### 4.5 Antioxidant activity

Venukumar and Latha found methanolic extract of *C. orchioides* has antioxidant activity in CCl4-induced hepatopathy in rats [73]. The methanolic extract was extremely effective in scavenging superoxide radicals. It expresses different levels of antioxidant activities in the experimental rats [74]. Methanolic extract of elevated rhizomes treated hepatotoxic rats reported increasing levels of antioxidant enzymes such as glutathione transferase [73]. Methanolic root extract showed DPPH (2,2-diphenyl-1-picryl-hydrazyl-hydrate) radical scavenging activity [15, 58]. Phenolic compounds were major components of *C. orchioides* expressing antioxidant activity reported [75]. Nitric oxide and DPPH radicals were effectively scavenged by using methanol extract of rhizome of *C. orchioides.* However, lipid peroxidation was only moderately effective [74]. Antioxidant activity of *C. orchioides* was confirmed by ferric reducing ability of plasma testing as well as 2,2′-azino-bis (3-ethylbenzothiazoline-6-sulfonic acid) [76]. To assess *C. orchioides'* antioxidant activity assays such as reducing power, DPPH, and phosphomolybdenum were used. Ethanolic extract of *C.*

*orchioides* root expressed significantly reducing power, anti-free radical, and antioxidant activity, compared with a reference standard, as gallic acid [77].

### 4.6 Anti-inflammatory activity

Mast cells of mice exposed to 48/80-induced systemic anaphylaxis were significantly inhibited from degranulating by extract of *C. orchioides*. The anti-inflammatory activity of *C. orchioides* ethanol extract was expressed to inhibitory effects on mice peritoneal mast cell degranulation and mast cell-derived immediate-type allergic reactions [78]. Carrageen-induced paw edema rats were significantly decreased by *C. orchioides* rhizome gel formulation [79]. Methanolic extract of rhizome of *C. orchioides* at a dose of 200 and 400 mg/kg expressed significant anti-inflammatory activity, compared with the standard drug, Diclofenac sodium (15 mg/kg) [80]. Ding et al. [81] found that curculigoside A of *C. orchioides* reduces significantly paw swelling and arthritis index in mice, decreases serum as IL-1β, TNF-α, IL-6, and PGE2 contents, increases serum superoxide dismutase activity in rats, induced with adjuvant arthritis. Curculigoside A of *C. orchioides* reduces serum pro-inflammatory factors, including IL-10, TNF-α, IL-17A, IL-6, IL-12, and IL-1β, and inhibits paw swelling in type II collagen-induced arthritis rats [82].

### 4.7 Antiosteoporosis activity

The rhizomes of *C. orchioides* were asserted to strengthen bones and tendons [83]. Curculigoside A (100 mM) enhances the proliferation of bone marrow stromal cells. The expression of osteogenic genes was enhanced, resulting in osteoprotegerin secretion being increased [84]. Cuculigoside A may inhibit production of the inflammatory cytokines such as TNF-a, IL-1a, IL-6, and COX-2 by dexamethasone-induced rat calvarial osteoblasts. At the same time, it regulates the expression of osteoblast COX-2, proliferation, and differentiation [85, 86].

The chlorophenolic glucosides including Curculigine M, Curculigine N, and Curculigine O were isolated from *C. orchioides* rhizomes. They expressed a moderate effect on osteoblast proliferation against the MC3T3-E1 cell line using MTT assays. COP70-3, a homogeneous heteropolysaccharide was isolated and purified from the crude polysaccharide (CO70) significantly improving the osteogenic mineralization rate and having favorable antiosteoporosis activity *in vitro* [51].

Curculigosides may decrease bone loss, induced by oxidative stress due to iron overload. Thus, curculigoside is used for the treatment of primary osteoporosis and bone loss in iron overload-related diseases [54, 55]. The major bioactive curculigosides of rhizomes, the phenolic glycoside exhibited neuroprotective and antiosteoporotic activities [12]. COP50-4, a polysaccharide isolated and purified from *C. orchioides* rhizome showed antiosteoporosis activity. It has great potential for the treatment of osteoporosis [87].

### 4.8 Antidepressant activity

Curculigosides significantly decreased NMDA-induced loss of neuron cells, production of reactive oxygen species, apoptosis, necrosis, and excitotoxicity in cultured cortical neurons. Curculigosides may inhibit the production of intracellular reactive oxygen species, and apoptosis in cultured cortical neurons. Curculigosides may prevent the dying of neurons and decrease their apoptosis and necrosis when exposed to neuronal excitotoxicity caused by NMDA [11]. Ge et al. [88] found that orcinols glucoside reduced activity of the hypothalamic-pituitary-adrenal axis, decreased depressive behaviors in chronic unpredictable mild stress rats by expressions of up-regulating BDNF and phosphorylating ERK1/2. Curculigoside reduced significantly as tests of the immobility time of the forced swimming, tail suspension, and osmotic fragility. Curculigosides increased significantly the contents of serotonin, dopamine, and norepinephrine and expressions of brain-derived neurotrophic factor (BDNF) protein in the hippocampus of chronic mild stress rats. Thus, curculigosides can be used for the treatment of depression in this way [89].

Extracts of *C. orchioides* rhizomes inhibited effectively neuroprotective activity of Acetylcholinesterase. It may be used for the treatment of Alzheimer's disease [90]. Assays of autobiography and spectrophotometry found the antiacetylcholinesterase activity of *C. orchioides* rhizome methanolic extract *in vitro*. It plays an important role in the treatment of Alzheimer's disease. It was explained by its inhibition effect of the acetylcholinesterase enzyme [91].

Leaf extracts of *C. orchioides* also contain 4-Acetyloxyimino-6,6-dimethyl-3-methylsulfanyl-4,5,6,7-tetrahydro benzo[c]thiophene-1 carboxylic acid methyl ester. It expressed the biological activities in the treatment of Alzheimer's, neurodegenerative diseases, and cognition disorders. Meanwhile, 2,7-Diphenyl-1,6-dioxopyridazino [4,5:2′,3′] pyrrolo[4′,5′-d] pyridazine of rhizome extract having biological activity as an antiepileptic, for treating renal disease, heart failure and Alzheimer's disease [19].

### 4.9 Hepatoprotective activity

Rao et al. [92] isolated curculignin A and curculigol, screened for their antihepatotoxic activity against thioacetamide and galactosomine-induced hepatoxic. They found the anti-inflammatory and hepatoprotective activities of *C. orchioides*. *C. orchioides* rhizome has hepatoprotective activity. Important enzyme activities such as gamma-glutamyl transpeptidase, aspartate aminotransferase, alkaline phosphatase, and gamma-glutamyl transpeptidases were significantly decreased in rats exposed to carbon tetrachloride. Increased food consumption and weight gain of mice were recorded due to the methanolic extract of *C. orchioides* rhizomes. In addition to lowering liver and serum protein levels, serum lipids, cholesterol, and phospholipids were normalized [73].

### 4.10 Nephroprotective activity

*C. orchioides* do not affect cyclophosphamide's chemotherapeutic efficacy when treatment of intestinal toxicities and hepatotoxicity, caused by cyclophosphamide. The extract of whole plant significantly reduced blood urea nitrogen, and serum creatinine levels in rats. Urotoxicity and nephrotoxicity induced by cyclophosphamide were also decreased [93].

### 4.11 Antidiabetic activity

The aqueous and ethanol extracts of *C. orchioides* are effective in being antihyperglycemic in diabetic rats induced by alloxan and loaded with glucose. Dose-dependent antihyperglycemic effect was observed after treatment with extract [94]. Ethanolic extract of *C. orchioides* may inhibit adipogenesis and enhance the uptake of glucose in 3 T3-L1 adipocytes [95]. The ethanolic and hydroalcoholic extracts of *C. orchioides* decreased changes in lipids caused by hyperglycemia, oxidative stress, and renal dysfunction in diabetic nephropathy rats induced by streptozotocin-nicotinamide [96].

### 4.12 Antimicrobial activity

There are several studies that investigated the insecticidal potential of *C. orchoisdes* extract against fungi and bacteria. It plays an important role in the treatment of diseases. Oil of *C. orchioides* rhizome can resist *Bacillus anthracis* and *B. subtilis.* At the same time, it was also found inhibiting *Cladosporium* species, *Salmonella* species *Fusarium solani*, *Staphylococcus aureus, Salmonella pullorum*, and *Aspergillus flavus, Fusarium moniliforme* [97]. In another study, the aqueous extract of *C. orchioides* displayed the antibacterial activity against Gram-negative bacteria such as *Salmonella typhimurium, Pseudomonas aeruginosa,* and *E. coli*), Gram-positive bacteria including *Staphylococcus epidermidis*, *S. aureus*). Thus, extracts of *C. orchioides* may be used for preventing effectively bacterial infections [16]. Otherwise, *C. orchioides* alcohol extract inhibited methicillin-resistant *P. aeruginosa* at a minimum inhibitory concentration value of 49 g/ml, whereas Gram-negative bacteria were not affected [98]. It is found that leaf extracts of *C. orchioides* carrying phytochemical-loaded silver nanoparticles, are strongly effective against *P. aeruginosa* and *S. aureus.* However, they are less effective against *E. coli* and *Klebsiella pneumonia* [99]. Yadav et al. [60] also investigated the alcoholic extract of *C. orchioides* having significant antimicrobial activities against bacterial pathogens and HepG2 cell lines.

### 4.13 Antiasthmatic activity

Pandit et al. [100] found that *C. orchioides* ethanolic extract relaxes goat tracheal chains and guinea pig ileum. It remarkably decreased passive paw anaphylaxis and bronchoconstriction in rats, guinea pigs, and mice with catalepsy induced by haloperidol. Thus, *C. orchioides* ethanolic extract has an antiasthmatic activity.

### 4.14 Antistress activity

Ethanolic extract of *C. orichioides* rhizome decreases immobility times at 200 mg/kg body weight in forced swimming and tail suspension tests. It enhanced mobility in tests, based on an actophotometer. Ethanolic extract of rhizomes increased tolerance to heat [50 and 70 infrared (IR) units] in IR testing and rotarod testing. It provided strong scientific evidence that the ethanolic extract of *C. orichioides* has antistress activity, used for potential antistress treatment [101].

### 4.15 Anticancer activity

Cancer diseases have caused human death across the globe. *C. orchoides* has been demonstrated to have potential anticancer effects against cancer diseases nowadays. One study found that the plant extract of *C. orchioides* when administered in combination with cyclophosphamide enhanced the anticancer activity of cyclophosphamide and ameliorated its toxic side effects [102]. In another study, silver nanoparticles carrying an extract of *C. orchioides* have efficacy against breast cancer cell lines in humans after 48 hours of incubation *in vitro* [103]. By enhancement of immune function and induction of apoptosis, polysaccharides of *C. orchioides* rhizome have a remarkably antitumor effect on cervical cancer *in vivo* and *in vitro* [104].

Furthermore, *C. orchioides* has also significant cytotoxicity activity. Both the aqueous extract of fresh root and the methanolic extract of the dried root are toxic to the human lung cancer cell line (adenocarcinoma NCI-H-522). Otherwise, Curculigines P compound of *C. orchioides* rhizomes expressed strongly the inhibitory activity on 5α-reductase enzyme by bioassay based on HaCaT. Thus, it may be used for prostatic hyperplasia treatment [52]. Hence, *C. orchioides* and active compounds derived from rhizomes can be used as useful potential medicines for the treatment and prevention of human cancer cell line development.

### 4.16 Antigout activity

Two heterocyclic phenolic derivatives, orcinosides I and J of *C. orchioides* rhizomes, displayed xanthine oxidase inhibitory activities with $IC_{50}$ values of 0.25 and 0.62 mM, respectively. Thus, these bioactivity compounds have an antigout effect, can used as medicines for gout disease treatment [48].

### 4.17 Antihypertensive activity

One study found that methanolic extract of *C. orchioides* root possesses antihypertensive activity by inhibiting angiotensin-converting enzyme in deoxycorticosterone acetate salt-induced hypertensive rats. It may be used against gout disease in humans [105].

### 4.18 Antimalarial activity

The study demonstrated that silver nanoparticles carrying extracts of *C. orchioides* have antimalarial activity with the highest mortality rate of malarial vectors as *Culex quinquefasciatus* and *Anopheles subpictus*, compared to control (without treatment of extract). So, extracts from C. orchioides can be used as antimalarial drugs [103].

### 4.19 Cardiovascular activity

The study demonstrated that curculigoside of *C. orchioides* may protect endothelial cells against $H_2O_2$-induced oxidative injury. It suggests that the curculigoside compound can be used as a promising drug for the treatment of cardiovascular disorders disease in humans [106].

## 5. Conclusion and future perspectives

*C. orchioides* is a native, precious medicinal plant, distributed in Vietnam and other subtropical and tropical countries of Asia. Due to over-exploitation from nature, while plant cultivation is still limited, it is in danger of extinction soon in Vietnam, India, China, and other countries in the world. The plants grow well on hills and mountains under the shade, where the soil is moist. Using analysis methods such as high-resolution spectroscopic, and high-performance liquid chromatography, researchers have isolated, purified, and determined the structure of compounds from the rhizomes and leaves. Research results have shown that *C. orchioides* contains several classes of chemicals such as saponins, alkaloids, flavonoids, phytosterols, steroids, free reducing sugars, starch, proteins, resins, tannins, mucilage, fats, calcium oxalate, inorganic compounds, and fibers. In which, many valuable compounds are identified such as curculigosides, orcinol glucosides, glucuronic acid, curculigol, curculignin P, polysaccharides. Alcoholic and methanolic extracts and some compounds derived from rhizomes and leaves of *C. orchioides* were used to evaluate biological effects on experimental animal models. Research results are scientific evidence, proving that they have many biological effects such as aphrodisiac, spermatogenic, estrogenic, immunomodulatory, antioxidant, anti-inflammatory, antiosteoporosis, antidepressant, hepatoprotective, nephroprotective, antidiabetic, antimicrobial, antiasthmatic, antistress, anticancer, antigout, antihypertensive, antimalarial and cardiovascular. In fact, Rhizoma Curculiginis has been used in traditional medicine remedies in Asia countries for a long time to treat diseases, including impotency and aphrodisiacs in men, tonics, jaundice, and skin diseases. The results of the biological activity of *C. orchioides* contributes to supporting using of traditional medicines and remedies to serve public health care.

Healthcare products derived from *C. orchioides* have been used in many Asian countries. They are used to treat infertility related to sperm, male erectile dysfunction, tonic for people with physical weakness, and many other diseases. However, until now there have not been many products derived from *C. orchioides* at commercial scale in the market. Because there is not much scientific evidence about its effects, as well as acute and chronic toxicity. It can be said that *C. orchioides* has a lot of potential for research and development of drugs that contribute to primary healthcare in the future. New discoveries of potential bioactive substances and their pharmacological effects will provide the possibility of safe drug applications. Therefore, there needs to be more and more research on *C. orchioides* such as cultivation, phytochemistry, pharmacology, preparation, and clinical trials.

## Author details

Nguyen Van Khiem*, Nguyen Minh Duc and Dinh Thanh Giang
National Institute of Medicinal Materials, Hanoi, Vietnam

*Address all correspondence to: ngvankhiem@yahoo.com

**IntechOpen**

© 2024 The Author(s). Licensee IntechOpen. This chapter is distributed under the terms of the Creative Commons Attribution License (http://creativecommons.org/licenses/by/4.0), which permits unrestricted use, distribution, and reproduction in any medium, provided the original work is properly cited.

## References

[1] Mishra SSB, Stierman B, Gahche JJ, Potischman N. Dietary supplement use among adults: United States, 2017-2018. NCHS Data Brief. 2021, 2021;**399**:1-8. U.S. Department of Health and Human Services, Centers for Disease Control and Prevention: Atlanta, GA, USA

[2] Zahn R, Perry N, Perry E, Mukaetova-Ladinska EB. Use of herbal medicines: Pilot survey of UK users' views. Complementary Therapies in Medicine. 2019;**44**:83-90

[3] Cecilia NC, Al Washali AY, Aisha A, Abuaisha A, Albishty MM, Suriani I, et al. The use of herbal medicine in Arab countries: A review. International Journal of Public Health and Clinical Sciences. 2017;**4**(5):1-14

[4] Insights FB. Herbal Medicine Market Size, Share & COVID-19 Impact Analysis, by Application (Pharmaceutical & Nutraceutical, Food & Beverages, and Personal Care & Beauty Products), Form (Powder, Liquid & Gel, and Tablets & Capsules) and Regional Forecast, 2021-2028. 2022. Available from: https://www.fortunebusinessinsights.com/herbal-medicine-market-106320 [Accessed: October 10, 2022]

[5] Ekor M. The growing use of herbal medicines: Issues relating to adverse reactions and challenges in monitoring safety. Frontiers in Pharmacology. 2014;**4**:177

[6] Singh R, Srivastava DK. Ethnobotanical studies on *Curculigo orchioides* from Sonebhadra District of Uttar Pradesh. Indian Journal of Science Research. 2014;**5**(2):123-124

[7] Loi DT. Vietnamese Medicinal Plants and Herbs. Hanoi: Medical Publishing House; 2004

[8] Chaturvedi P, Briganza V. Enhanced synthesis of curculigoside by stress and amino acids in static culture of *Curculigo orchioides* Gaertn (Kali Musli). Pharmacognosy Research. 2016;**8**(3):193-198

[9] Nie Y, Dong X, He Y, Yuan T, Han T, Rahman K, et al. Medicinal plants of genus *Curculigo*: Traditional uses and a phytochemical and ethnopharmacological review. Journal of Ethnopharmacology. 2013;**147**(3):547-563

[10] Wang Y, Li J, Li N. Phytochemistry and pharmacological activity of plants of genus *Curculigo*: An updated review since 2013. Molecules. 2021;**26**(11):3396

[11] Tian Z, Yu W, Liu HB, Zhang N, Li XB, Zhao MG, et al. Neuroprotective effects of curculigoside against NMDA-induced neuronal excitoxicity *in vitro*. Food and Chemical Toxicology. 2012;**50**(11):4010-4015

[12] Zhu F, Wang J, Ni Y, Yin W, Hou Q, Zhang Y, et al. Curculigoside protects against titanium particle-induced osteolysis through the enhancement of osteoblast differentiation and reduction of osteoclast formation. Journal of Immunology Research. 2021:5707242

[13] Agrahari AK, Panda SK, Meher A, Padhan AR. Studies on the anti-inflammatory properties of *Curculigo orchioides* Gaertn. Root tubers. International Journal of Pharmaceutical Sciences and Research. 2010;**1**:139-143

[14] Madhavan V, Joshi R, Murali A, Yoganarasimhan SN. Antidiabetic activity of *Curculigo orchioides*. Root Tuber. Pharmaceutical Biology. 2008;**45**:18-21

[15] Kushalan S, Yathisha UG, Khyahrii SA, Hegde S. Phytochemical and antioxidant evaluation of *in vitro* and *in vivo* propagated plants of *Curculigo orchioides*. In Vitro Cellular & Developmental Biology – Plant. 2022;**58**(3):382-391

[16] Nagesh KS, Shanthamma C. Antibacterial activity of *Curculigo orchioides* rhizome extract on pathogenic bacteria. African Journal of Microbiology Research. 2009;**3**(1):5-9

[17] Chopra RN, Nayar SL, Chopra IC. Glossary of Indian medicinal plants. The Quarterly Review of Biology. 1956;**33**:330

[18] Ministry of Science and Technology, and Vietnam Academy of Science and Technology. Vietnam Red Data Book. Part II. Plants. Hanoi, Vietnam: Science and Technology Publishing House; 2007

[19] Brintha S, Rajesh S, Renuka R, Santhanakrishnan VP, Gnanam R. Phytochemical analysis and bioactivity prediction of compounds in methanolic extracts of *Curculigo orchioides* Gaertn. Journal of Pharmacognosy and Phytochemistry. 2017;**6**(4):192-197

[20] Joy PP. Agrotechnological practices for quality crude drug production in Nilappana (*Curculigo orchioides* Gaertn) [thesis]. India: Kerala Agricultural University; 2003

[21] Joy PP, Savithri KE, Mathew S, Skaria BP. Growth yield and quality of two biotypes of black Musli (*Curculigo orchioideds*). Journal of Medicinal and Aromatic Plant Sciences. 2004;**26**(4):684-692

[22] Joy PP, Savithri KE. Effect of mulch and sources of nutrients on growth, yield and quality of black musli (*Curculigo orchioides*). Journal of Medicinal and Aromatic Plant Sciences. 2005;**27**:646-656

[23] Singh T, Singh AP, Vyas M, Khurana N, Sharma N, Kumar R, et al. A retrospective study on cultivation, collection and pharmacological activities of endangered plant-"*Curculigo orchoides*". Plant Cell Biotechnology and Molecular Biology. 2021;**22**(33&34):185-196

[24] Joy PP, Thomas J, Samuel M, Skaria BP. *Curculigo orchioides*: A plant for health care. Indian Journal of Arecanut, Spices and Medicinal Plants. 2004;**6**(4):131-134

[25] Raghunathan K, Mitra R. Pharmacognosy of Indigenous Drugs-Vol. I. New Delhi: Central Council for Research in Ayurveda and Siddha; 1982. pp. 127-136

[26] Asif M. A review on phytochemical and ethnopharmacological activities of *Curculigo orchioides*. Mahidol University Journal of Pharmaceutical Sciences. 2012;**39**(3-4):1-10

[27] Le VTT, Dung NT, Dai TD, Hien NT. Chemical composition and biological activities of *Curculigo orchioides* growing in Tuyen Quang province. TNU Journal of Science and Technology. 2022;**227**(08):527-533

[28] Mehta BK, Bokadia MM, Mehta SC. Study of root oil compound fatty acids of *Curculigo orchioides* roots. Indian Drugs. 1980;**18**(3):109-110

[29] Kubo M, Namba K, Nagamoto N, Nagao T, Nakanishi J, Uno H, et al. New phenolic glucoside, curculigoside from the rhizomes of *Curculigo orchioides*. Planta Medica. 1983;**47**(1):52-55

[30] Misra TN, Singh RS, Tripathi DM, Sharma SC. Curculigol, a cycloartane

triterpene alcohol from *Curculigo orchioides*. Phytochemistry. 1990;**29**(3):929-931

[31] Shrama PC, Yelne MB, Dennis TJ. Database on Medicinal Plants Used in Ayurveda. Vol. 4. New Delhi: Ministry of Health & Family Welfare; 2002

[32] Chaterjee A, Prakashi SC. The Treatise on Indian Medicinal Plants. Vol. 6. New Delhi: NISCAIR; 2001

[33] Misra TN, Singh RS, Upadhyay J, Tripathi DNM. Aliphatic hydroxy ketones from *Curculigo orchioides* rhizome. Phytochemistry. 1984;**23**(8):1643-1645

[34] Misra TN, Singh RS, Tripathi DM. Aliphatic compounds from *Curculigo orchioides* rizhomes. Phytochemistry. 1984;**23**(10):2369-2371

[35] Porwal M, Batra A, Mehta BK. Some new compounds from the rhizome of *Curculigo orchioides* Gaertn. Indian Journal of Chemistry. 1988;**27 B**:856-857

[36] Porwal M, Mehta BK. *Curculigo orchioides*: A medicinally important plant. Nagarjun. 1985;**29**(3):12-13

[37] Garg SN, Misra LN, Agarwal SK. Corchioside A: An orcinol glycoside from *Curculigo orchioides*. Phytochemistry. 1989;**28**(6):1771-1772

[38] Xu JP, Xu RS, Yuli X. Glycosides of cycloartene sapogenine from *Curculigo orchioides*. Phytochemistry. 1992;**31**:223-236

[39] Xu JP, Xu RS. Cycloartane-type sapogenins and their glycosides from *Curculigo orchioides*. Phytochemistry. 1992;**31**(7):2455-2458

[40] Xu JP, Xu RS, Li XY. Four new cycloartane saponins from *Curculigo orchioides*. Planta Medica. 1992;**58**(2):208-210

[41] Goyal P, Kabra A. A review on phytochemical and pharmacological profile on *Curculigo orchioides*. Plant Cell Biotechnology and Molecular Biology. 2020;**21**:243-252

[42] Saxena HO. Phytochemical screening and variation studies in secondary metabolite. Chemical Science Review and Letters. 2022;**11**(42):151-158

[43] Wang X, Zhang M, Zhang D, Wang S, Yan C. An O-acetylglucomannan from the rhizomes of *Curculigo orchioides*: Structural characterization and anti-osteoporosis activity *in vitro*. Carbohydrate Polymers. 2017;**174**:48-56

[44] Wang X, Zhang M, Zhang D, Wang X, Cao H, Zhang Q, et al. Structural elucidation and anti-osteoporosis activities of polysaccharides obtained from *Curculigo orchioides*. Carbohydrate Polymers. 2019;**203**:292-301

[45] Wang ZH, Huang J, Ma XC, Li GY, Ma YP, Li N, et al. Phenolic glycosides from *Curculigo orchioides* Gaertn. Fitoterapia. 2013;**86**:64-69

[46] Wang ZH, Ma XC, Li GY, Niu C, Ma YP, Kasimu R, et al. Four new phenolic glucosides from *Curculigo orchioides* Gaertn. Phytochemistry Letters. 2014;**9**:153-157

[47] Chen X, Zuo A, Deng Z, Huang X, Zhang X, Geng C, et al. New phenolic glycosides from *Curculigo orchioides* and their xanthine oxidase inhibitory activities. Fitoterapia. 2017;**122**:144-149

[48] Niu C, Zhang ZZ, Yang LP, Zhai YY, Wang ZH. Chemical constituents of *Curculigo orchioides*. Chemistry of Natural Compounds. 2020;**56**:1-3

[49] Cao DP, Han T, Zheng YN, Qin LP, Zhang QY. Phenolic glycosides and lignans components in *Curculigo orchioides* Gaertn. Academic journal of Second Military Medical University. 2009;**30**:194-197

[50] Xu JP, Dong QY. Chemical study on *Curculigo orchioides* II, the isolation and structural determination of new compound curculigine A. Chinese Traditional and Herbal Drugs. 1987;**18**:194-196. [in Chinese]

[51] Wang ZH, Gong XY, Zhou DJ, Xu PF, Huang M, Zhang QL, et al. Three new chloro-phenolic glucosides from *Curculigo orchioides* Gaertn. Phytochemistry Letters. 2018;**26**:9-11

[52] Deng XL, Zheng RR, Han ZZ, Gu LH, Wang ZT. New chlorophenolic glycoside from *Curculigo orchioides* and their activities on 5α-reductase. Journal of Asian Natural Products Research. 2021;**23**(4):333-340

[53] Jiao W, Chen X, Wang H, Lu R, Shao H. A new hepatotoxic triterpenoid ketone from *Curculigo orchioides*. Fitoterapia. 2013;**84**:1-5

[54] Zhang J, Li YD, Liu XM, Gao L, Zhang YP, Dang LZ. The study of terpeniods from *Curculigo orchioides*. Journal of Yunnan University. 2019;**41**:367-371

[55] Zhang QL, Zhao L, Shen Y, He YQ, Cheng G, Yin M, et al. Curculigoside protects against excess-iron induced bone loss by attenuating Akt-FoxO1-dependent oxidative damage to mice and osteoblastic MC3T3-E1 cells. Oxidative Medicine and Cellular Longevity. 2019(8):1-14

[56] Umar AH, Ratnadewi D, Rafi M, Sulistyaningsih YC. Untargeted metabolomics analysis using FTIR and UHPLC-Q-Orbitrap HRMS of two *Curculigo* species and evaluation of their antioxidant and alpha-glucosidase inhibitory activities. Metabolites. 2021;**11**:42

[57] Kushalan S, D'Souza LC, Aloysius K, Sharma A, Hegde S. Toxicity assessment of *Curculigo orchioides* leaf extract using *Drosophila melanogaster*: A preliminary study. International Journal of Environmental Research and Public Health. 2022;**19**(22):15218

[58] Ngoc NB, Tuyen NM, Thuong PT, Van Tuan V, Van Khiem N, Hong VD. Three phenolic glycoside compounds isolated from the rhizome of *Curculigo orchioides* Gaertn. Collected in the central highlands. Journal of Pharmacology. 2015;**55**(7):54-57

[59] Pham T, Phan N, Tran VG, Nguyen TQ, Tran MD, Nguyen TH. Analysis of a major phenolic glucoside and biochemistry compounds from *Curculigo orchioides* Gaertn. Reports of Biochemistry And Molecular Biology. 2022;**10**:565-572

[60] Yadav D, Shukla S. Mass spectrometric imaging of rhizomes of *Curculigo orchioides* and their bioactivities. Toxicology Open Access. 2017;**3**(Suppl 1). DOI: 10.4172/2476-2067.C1.002

[61] Chauhan NS, Rao CV, Dixit VK. Effect of *Curculigo orchioides* rhizomes on sexual behaviour of male rats. Fitoterapia. 2007;**78**(7-8):530-534

[62] Chauhan NS, Dixit VK. Spermatogenic activity of rhizomes of *Curculigo orchioides* Gaertn in male rats. International Journal of Applied Research in Natural Products. 2008;**1**(2):26-31

[63] Bui-Le T-N, Hoang-Tan Q, Hoang-Viet H, Truong-Thi B-P,

Nguyen-Thanh T. Protective effect of *Curculigo orchioides* Gaertn. Extract on heat stress-induced spermatogenesis complications in murine model. Current Issues in Molecular Biology. 2023;**45**:3255-3267

[64] Thakur M, Chauhan NS, Sharma V, Dixit VK, Bhargava S. Effect of *Curculigo orchioides* on hyperglycemia-induced oligospermia and sexual dysfunction in male rats. International Journal of Impotence Research. 2012;**24**:31-37

[65] Cao S, Tian S, Bai M, Liu SY, Jia JJ, Miao M. Effects of *Curculigo orchioides* total glucosides in mouse perimenopause model of related organization and organs morphology. Bangladesh Journal of Pharmacology. 2016;**11**:S72-S81

[66] Miao M, Tian S, Guo L, Bai M, Fang X, Liu S. The effect of curculigoside on mouse model of perimenopausal depression. Saudi Journal of Biological Sciences. 2017;**24**:1894-1190

[67] Chen QS, Chen WQ, Yang SY. Pharmacologic study of *Curculigo orchioides* Gaertn. Zhongguo Zhong Yao Za Zhi. 1989;**14**:618-620

[68] Lakshmi V, Pandey K, Puri A, Saxena RP, Saxena KC. Immunostimulant principles from *Curculigo orchioides*. Journal of Ethnopharmacology. 2003;**89**:181-184

[69] Zhou Y, Zhang L, Zhao LY, Zhang GY, Ma XQ, Ge DY, et al. Experimental study of immunoregulative actions of xianmaopolysaccharid (XMPS) in mice. Shanghai Journal of Immunology. 1996;**16**:336-338

[70] Bafna AR, Mishra SH. Immunostimulatory effect of methanol extract of *Curculigo orchioides* on immunosuppressed mice. Journal of Ethnopharmacology. 2006;**104**:1-4

[71] Lacaille-Dubois MA, Wagner H. A review of the biological and pharmacological activities of saponins. Phytomedicine. 1996;**2**:363-386

[72] Ji XH. Effect of *Curculigo* polysaccharide on immune function mice. Hai Xia Yao Xue. 2011;**23**:33-35. [in Chinese]

[73] Venukumar MR, Latha MS. Antioxidant activity of *Curculigo orchioides* in carbon tetrachloride induced hepatopathy in rats. Indian Journal of Clinical Biochemistry. 2012;**17**(2):80-87

[74] Bafna AR, Mishra SH. *In vitro* antioxidant activity of methanol extract of rhizomes of *Curculigo orchioides* Gaertn. ARS Pharma. 2005;**46**(2):125-138

[75] Wu Q, Fu DX, Hou AJ, Lei GQ, Liu ZJ, Chen JK, et al. Antioxidative phenols and phenolic glycosides from *Curculigo orchioides*. Chemical & Pharmaceutical Bulletin. 2005;**53**:1065-1067

[76] Surveswaran S, Cai YZ, Corke H, Sun M. Systematic evaluation of natural phenolic antioxidants from 133 Indian medicinal plants. Food Chemistry. 2007;**102**:938-953

[77] Ratnam KV, Ravishankar K, Priyabhandavi P. Evaluation of *in vitro* antioxidant activity of ethanolic root extract of *Curculigo orchioides*. International Journal of Pharmaceutical Chemistry. 2013;**3**(2):364-369

[78] Venkatesh P, Mukherjee PK, Nema NK, Bandyopadhyay A, Fukui H, Mizuguchi H. Mast cell stabilization and antihistaminic potentials of *Curculigo orchioides* rhizomes. Journal of Ethnopharmacology. 2009;**126**(3):434-436

[79] Dode PA, Wani NS, Deshmukh TA, Patil VR. Anti-inflammatory activity of hydrogel formulations of *Curculigo orchioides* Gaertn. Rhizomes. Pharmacology. 2009;**2**:1367-1381

[80] Mohammad A, Kumar A. Acute toxicity study and *In-vivo* anti-inflammatory activity of different fractions of *Curculigo orchioides* Gaertn. Rhizome in albino Wistar rats. Iranian Journal of Pharmaceutical Sciences. 2010;**6**(3):191-198

[81] Ding H, Gao G, Zhang L, Shen G, Sun W, Gu Z, et al. The protective effects of curculigoside a on adjuvant-induced arthritis by inhibiting NF-κB/NLRP3 activation in rats. International Immunopharmacology. 2016;**30**:43-49

[82] Tan S, Xu J, Lai A, Cui R, Bai R, Li S, et al. Curculigoside exerts significant anti-arthritic effects *in vivo* and *in vitro via* regulation of the JAK/STAT/NF-κB signaling pathway. Molecular Medicine Reports. 2019;**19**:2057-2064

[83] Cao D, Zheng Y, Qin L, Han T, Zhang H, Rahman K, et al. *Curculigo orchioides*, a traditional Chinese medicinal plant, prevents bone loss in ovariectomized rats. Maturitas. 2008;**59**:373-380

[84] Shen QP, Zeng DL, Zhou YT, Xia LG, Zhao YF, Qiao GY, et al. Curculigoside promotes osteogenic differentiation of bone marrow stromal cells from ovariectomized rats. The Journal of Pharmacy and Pharmacology. 2013;**65**:1005-1013

[85] Zhu FB, Wang JY, Zhang YL, Quan RF, Yue ZS, Zeng LR, et al. Curculigoside regulates proliferation, differentiation, and pro-inflammatory cytokines levels in dexamethasone-induced rat calvarial osteoblasts. International Journal of Clinical and Experimental Medicine. 2015;**8**:12337-12346

[86] Zhu H, He J, Ye L, Lin F, Hou J, Zhong Y, et al. Mechanisms of angiogenesis in a Curculigoside A-treated rat model of cerebral ischemia and reperfusion injury. Toxicology and Applied Pharmacology. 2015;**288**(3):313-321

[87] Yu Y, Li T, Wang X, Zhang M, Yu Q, Chen H, et al. Structural characterization and anti-osteoporosis activity of two polysaccharides extracted from the rhizome of *Curculigo orchioides*. Food & Function. 2022;**13**(12):6749-6761

[88] Ge JF, Gao WC, Cheng WM, Lu WL, Tang J, Peng L, et al. Orcinol glucoside produces antidepressant effects by blocking the behavioural and neuronal deficits caused by chronic stress. European Neuropsychopharmacology. 2014;**24**(1):172-180

[89] Wang J, Zhao XL, Gao L. Anti-depressant-like effect of curculigoside isolated from *Curculigo orchioides* Gaertn root. Tropical Journal of Pharmaceutical Research. 2016;**15**(10):2165-2172

[90] Pratap GK, Shantaram M. A kinetic study of acetylcholinesterase inhibition by fractions of *Oleo dioica* Roxb. Leaf and *Curculigo orchioides* Gaertn rhizome for the treatment of Alzheimer's disease. EJMP. 2019;**30**:1-12

[91] Pratap GK. *In vitro* anti-cholinesterase activity and mass spectrometric analysis of *Curculigo orchioides* Gaertn. Rhizome extract. Analytical Chemistry Letters. 2020;**10**:442-458

[92] Rao KS, Mishra SH. Studies on *Curculigo orchioides* Gaertn for anti-inflammatory and hepatoprotective activities. Indian Drugs. 1996;**33**(1):20-25

[93] Murali VP, Kuttan G. *Curculigo orchioides* Gaertn effectively ameliorates the uro-and nephrotoxicities induced by cyclophosphamide administration in experimental animals. Integrative Cancer Therapies. 2016;**15**(2):205-215

[94] Madhavan V, Joshi R, Murali A, Yoganarasimhan SN. Antidiabetic activity of *Curculigo orchioides* root tuber. Pharmaceutical Biology. 2007;**45**(1):18-21

[95] Gulati V, Gulati P, Harding IH, Palombo EA. Exploring the anti-diabetic potential of Australian aboriginal and Indian ayurvedic plant extracts using cell-based assays. BMC Complementary and Alternative Medicine. 2015;**15**:8

[96] Singla K, Singh R. Nephroprotective effect of *Curculigo orchiodies* in streptozotocin–nicotinamide induced diabetic nephropathy in wistar rats. Journal of Ayurveda and Integrative Medicine. 2020;**11**:399-404

[97] Jaiswa S, Batra A, Mehta BK. The antimicrobial efficiency of root oil against human pathogenic bacteria and phytopathogenic fungi. Phytopathology. 1984;**109**:90-93

[98] Marasini BP, Baral P, Aryal P, Ghimire KR, Sanjiv N, Nabaraj D, et al. Evaluation of antibacterial activity of some traditionally used medicinal plants against human pathogenic bacteria. BioMed Research International. 2015:265425

[99] Perumal V, Thamilchelvan K, Jinu U, Giovanni B, Natesan G. Enhanced antibacterial and cytotoxic activity of phyto-chemical loaded silver nanoparticles using *Curculigo orchioides* leaf extracts with different extraction techniques. Journal of Cluster Science. 2017;**28**:607-619

[100] Pandit P, Singh A, Bafna AR, Kadam PV, Patil MJ (2008) Evaluation of antiasthmatic activity of *Curculigo orchioides* Gaertn. Rhizomes. Indian Journal of Pharmaceutical Sciences. 2008;**70**(4):440-444

[101] Chauhan NS, Shah K, Gupta PK. Studies on antistress activity of *Curculigo orchioides* Gaertn. Biomedical and Biotechnology Research Journal. 2021;**5**(2):145-148

[102] Murali VP, Kuttan G. Enhancement of cancer chemotherapeutic efficacy of cyclophosphamide by *Curculigo orchioides* Gaertn and its ameliorative effects on cyclophosphamide-induced oxidative stress. Integrative Cancer Therapies. 2015;**14**(2):172-183

[103] Kayalvizhi T, Ravikumar S, Venkatachalam P. Green synthesis of metallic silver nanoparticles using *Curculigo orchioides* rhizome extracts and evaluation of its antibacterial, larvicidal, and anticancer activity. Journal of Environmental Engineering. 2016;**142**(09):C401600237

[104] Xia LF, Liang SH, Wen H, Tang J, Huang Y, Wen H. Anti-tumor effect of polysaccharides from rhizome of *Curculigo orchioides* Gaertn on cervical cancer. Tropical Journal of Pharmaceutical Research. 2016;**15**(8):1731-1737

[105] Joshi UH, Solanki VR, Desai TR, Tirgar PR. Investigation of antihypertensive mechanism of *Curculigo orchioides* in DOCA salt induced hypertensive rats. International Journal of Phytopharmacology. 2012;**3**(2):178-185

[106] Wang YK, Hong YJ, Wei M, Wu Y, Huang ZQ, Chen RZ, et al. Curculigoside attenuates human umbilical vein endothelial cell injury induced by $H_2O_2$. Journal of Ethnopharmacology. 2010;**132**(1):233-239

**Chapter 7**

# Recent Advances in Phytochemical Research for Cancer Treatment

*Dexter Achu Mosoh*

## Abstract

Cancer is a significant global health concern, responsible for 16.8% of deaths and 22.8% of noncommunicable diseases. Currently, cancer ranks as the second leading cause of mortality, with projections indicating its rise to the top by 2060. The genomic, transcriptomic, and proteomic diversity of cancer cells contributes to their increased resistance to treatment, enhanced survival, and propensity for metastasis. Conventional cancer treatments like surgery, chemotherapy, and radiation often result in significant damage to healthy tissues. Phytochemicals, naturally occurring plant compounds, have gained attention as promising cancer therapies due to their anticancer, antioxidant, and anti-inflammatory properties. This chapter delves into recent advances in phytochemical research, emphasizing molecular mechanisms and innovative approaches for incorporating these compounds into cancer treatment. Additionally, it explores the challenges and opportunities of leveraging phytochemicals to bridge traditional medicine and modern oncology, offering hope for enhanced treatment outcomes and improved patient quality of life.

**Keywords:** computer-aided drug design (CADD), cancer, integrative oncology, medicinal plants, molecular mechanisms, multi-drug resistance, nano-phytomedicine, nanotechnology, natural product applied science, pharmacology, phytochemicals, phytochemical-based cancer management, plant extracts, signal transduction pathways, targeted phytotherapy, tumorigenesis

## 1. Introduction

Cancer remains a significant global health challenge, with staggering statistics highlighting its impact. Globally, cancer accounts for nearly one-sixth (16.8%) of all fatalities and nearly one-quarter (22.8%) of all deaths from noncommunicable diseases (NCDs) [1]. In the majority of nations (177 out of 183), cancer ranks among the top three causes of mortality for individuals aged 30–69, contributing to 30.3% of premature deaths from NCDs worldwide [1]. Reports from the World Health Organization and the American Cancer Society underscore the immense burden of cancer, with it accounting for the highest clinical, social, and economic burden among all human diseases, as measured by cause-specific disability-adjusted life years (DALYs) [2]. Alarmingly, the risk of developing cancer stands at 20.2% for men and

18.2% for women aged 0–74. In 2018 alone, there were 18 million new cancer cases diagnosed, with lung, breast, and prostate cancers being the most prevalent [2]. Projections indicate a staggering 70% increase in new cancer cases and deaths over the next two decades, underscoring the urgent need for advancements in cancer epidemiology and related research to combat this escalating public health crisis effectively [3].

Globally, cancer ranks second in terms of mortality, surpassed only by ischemic heart disease; however, projections suggest that by 2060, cancer will surpass all other causes of death [2]. Among the most lethal forms of cancer are those affecting the lungs, liver, and stomach. Notably, lung and breast cancers stand out as the leading causes of cancer-related deaths among both women and men [2]. In 2010, the United States allocated over $125 billion to cancer-related expenses. This figure is projected to exceed $245 billion by 2030 [4]. In 2024 alone, approximately 611,720 Americans are expected to die from cancer [5]. Lung and bronchus cancer are anticipated to be the leading causes of cancer mortality, accounting for around 125,070 deaths.

Cancer cells exhibit diverse genomic, transcriptomic, and proteomic profiles, which confer various phenotypic traits, including increased treatment resistance, enhanced survival, and a heightened propensity for metastasis [6]. This phenomenon, known as cancer heterogeneity, significantly complicates treatment and remains one of the foremost challenges in contemporary medicine [6–8]. As a result, cancer treatment is a complex and continuously evolving field, requiring innovative approaches and personalized therapies to effectively combat the diverse manifestations of the disease. Recent advances in cancer therapy, such as surgery, chemotherapy, radiation therapy, and targeted therapies, have greatly improved survival rates and quality of life for many patients [9, 10]. These improvements are especially noticeable in the fields of immunotherapy and targeted therapy, which have become novel pillars of cancer care [11]. The advancement of personalized medicine and the use of advanced technologies, such as next-generation sequencing, have further improved the effectiveness of these treatments [11]. However, cancer remains a leading cause of morbidity and mortality worldwide, necessitating ongoing advancements in its treatment.

Despite significant progress in medical science, the fight against cancer presents numerous challenges and treatments are not without significant limitations. Traditional cancer therapies, such as surgery, chemotherapy, and radiation therapy, often fail to target cancer cells exclusively, resulting in collateral damage to healthy tissues [12, 13]. For instance, chemotherapy, a widely used cancer treatment, often results in severe side effects such as fatigue, nausea, and increased susceptibility to infections, significantly diminishing a patient's quality of life [14–16]. This is due to the indiscriminate nature of chemotherapy, which targets not only cancer but also healthy cells and tissues [16, 17]. Additionally, the development of resistance in cancer cells can further undermine the efficacy of chemotherapy, leading to relapse and metastasis [16]. This issue arises from the reductionist approach to drug development and the intricate genetic mutations present in solid tumors [13]. While using molecular biomarkers can enhance the success of targeted therapies, many clinical trials proceed without incorporating these biomarkers, leading to missed therapeutic targets [18]. Nanocarrier-based cancer therapies hold promise for delivering chemotherapeutics precisely to cancer cells; however, challenges with nonselective targeting persist [19]. Additionally, researchers are actively exploring cellular therapies that leverage the immune system, but they face obstacles such as the immune-suppressive tumor microenvironment and potential off-target toxicity [20]. Addressing these challenges is crucial for advancing the efficacy and specificity of cancer treatments.

The complexity of cancer, driven by its genetic and molecular heterogeneity, necessitates personalized treatment approaches. For instance, previous research emphasizes the critical importance of understanding underlying biological mechanisms and acquiring molecular data to tailor therapies to individual patients [21, 22]. Another study further underscores the necessity of in-depth molecular analysis, including comprehensive molecular tumor profiling, to predict therapeutic responses accurately [23]. A previous study applied these principles to colorectal cancer, emphasizing the significant roles of genetic and epigenetic factors in disease heterogeneity and the potential of personalized medicine to optimize treatment selection [24]. Collectively, these studies highlight the imperative for a data-driven, personalized approach to cancer treatment, focusing on the unique molecular characteristics of each patient's tumor to improve outcomes. However, personalized medicines encounter hurdles in securing regulatory approval and garnering widespread acceptance among healthcare stakeholders [25]. They are required to demonstrate superior efficacy compared to conventional medical approaches, especially given the often high costs associated with tailored therapies [25]. While molecular profiling (MP)-guided therapy holds promise in cancer treatment, findings from the SHIVA trial revealed no significant improvement in progression-free survival, underscoring the need for further research and refinement in this area [26].

Personalized medicine in cancer treatment utilizes genetic, immunological, and proteomic profiling to provide tailored therapeutic options and prognostic insights. In contrast, targeted therapies, employed by researchers and clinicians, customize treatment based on the molecular characterization of malignant cancers [27]. Targeted therapies have significantly transformed cancer treatment, yet their impact is constrained by the emergence of resistance mechanisms and the genetic variability inherent in tumors [28, 29]. Various factors, such as genetic mutations and epigenetic modifications, can cause resistance [30]. Overcoming these hurdles requires a more precise approach to targeted therapies, which could involve using combination treatments that target multiple pathways or resistance mechanisms simultaneously [31, 32].

Immunotherapy has significantly transformed cancer treatment paradigms by leveraging the body's immune system to fight cancer cells [33]. Despite its groundbreaking success in specific cancer types, its effectiveness remains limited in many others, accompanied by the potential for severe immune-related adverse events [34]. Nevertheless, the remarkable achievements of immunotherapy have solidified its position as a cornerstone of cancer management, as highlighted by over 50 FDA approvals across various cancer indications [35]. Yet, significant challenges persist, notably the absence of robust predictive biomarkers and a comprehensive understanding of resistance mechanisms and the causality behind immune-related adverse events [35]. Timely identification and management of immune-related adverse events are paramount [34].

While recent strides in targeted therapies and immunotherapies have indeed transformed cancer treatment, they are not without hurdles. Challenges persist, including the unpredictable nature of treatment efficacy and patient response, the necessity for enhanced biomarkers, and the emergence of resistance to these therapies [36]. Moreover, the considerable costs and restricted accessibility of these treatments exacerbate these obstacles [36]. Although endeavors to enhance the accessibility and affordability of such therapies, notably in cellular immunotherapy, are underway, their broad adoption hinges on factors such as specificity, safety, and cost-effectiveness [37, 38]. The constraints inherent in existing treatments underscore the urgency for novel therapeutic approaches and a comprehensive grasp of cancer biology to forge tailored, efficacious treatments with reduced adverse effects. Elevating patient outcomes and bolstering survival rates remain paramount objectives in oncological research, with a pivotal emphasis on creating treatments that are more precise, potent, and less burdensome in toxicity.

## 2. Overview of cancer development

Cancer is a complex disease that progresses through several stages: initiation, promotion, conversion, progression, invasion, and metastasis (**Figure 1**) [39]. Chemotherapy primarily targets advanced stages of cancer but has limitations such as severe side effects, high costs, and a narrow focus on single-target mechanisms. Many cancers require multi-targeted approaches, like colorectal cancer, which often needs multiple therapeutic strategies [40]. Phytochemicals present a compelling alternative to conventional cancer therapies, providing enhanced therapeutic efficacy, cost-effectiveness, and reduced adverse effects. These naturally occurring compounds have the potential to impact multiple stages of cancer development, offering a multifaceted approach to both treatment and prevention [41]. Integrating phytochemicals into cancer therapy can lead to more targeted and effective interventions, ultimately improving patient outcomes and advancing sustainable cancer management strategies.

Carcinogens can act as initiators or promoters, impacting gene expression and leading to the activation of oncogenes or the suppression of tumor suppressor genes, which promote tumor development. Mutations can arise from various sources, including DNA damage from infections, UV-induced tissue injury, reactive oxygen species, and DNA adducts. Epigenetic modifications and the cellular microenvironment are key factors in cancer development. Inappropriate inflammatory responses can influence various stages of cancer. Pro-inflammatory components can drive

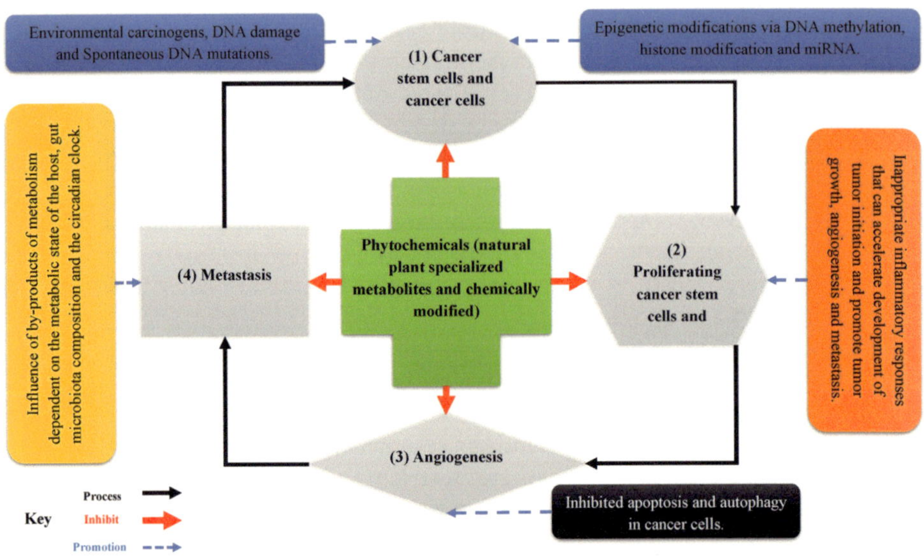


**Figure 1.**
*Schematic representation of cancer development and potential strategies for cancer prevention and therapy through phytochemical intervention.*

*DOI: http://dx.doi.org/10.5772/intechopen.1007200*

tumor initiation and progression, while anti-inflammatory components can support growth, angiogenesis, and metastasis. Metabolic by-products, the host's metabolic state, gut microbiota composition, and circadian rhythms collectively shape the cellular microenvironment [39]. Metabolic diseases and shifts in the gut microbiota, linked to disruptions in circadian alignment, can promote tumorigenesis. Elevated levels of insulin (hyperinsulinemia) and glucose (hyperglycemia) accelerate tumor growth, while conditions such as fatty liver and cholestasis interfere with the circadian rhythms of peripheral tissues, potentially exacerbating cancer progression (**Figure 1**).

## 3. Phytochemicals: Emerging promises for cancer therapy

Plant-based phytochemicals and their derivatives hold tremendous promise for enhancing the efficacy of cancer treatment, thanks to their diverse biological activities and propensity to mitigate adverse effects [42–44]. Inherently found in nature, these bioactive compounds represent a rich resource of potential anticancer agents. These compounds, spanning alkaloids, flavonoids, terpenoids, and phenolic compounds, have undergone extensive scrutiny for their anticancer potential (**Figure 2**) [43, 45–48]. Noteworthy examples include taxol and podophyllotoxin analogs, which exhibit the ability to modulate molecular pathways crucial for cancer proliferation and progression [42, 49, 50]. Their distinct advantages over conventional treatments, including safety, potency, and cost-effectiveness, have propelled them into focus [51]. Encouragingly, compounds like sulforaphane and vinca alkaloids have demonstrated promise in both preclinical and clinical settings [45]. In addressing skin cancer, phytochemicals have shown antioxidant, anti-proliferative, and anti-inflammatory properties, thus presenting potential alternatives or complements to traditional

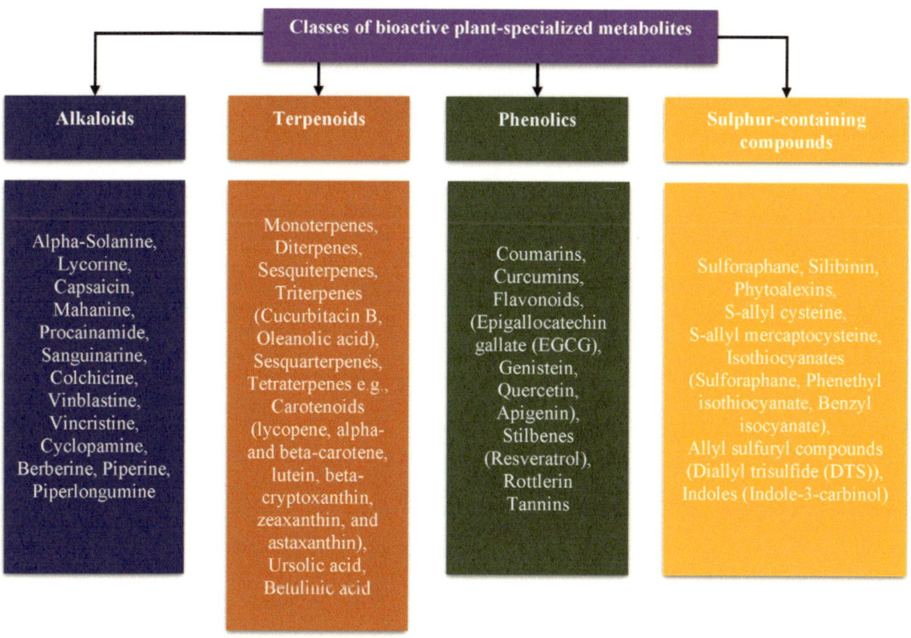


**Figure 2.**
*Overview of various classes of natural phytochemicals with potential roles in cancer chemoprevention.*

therapies [52]. Moreover, they wield the capacity to reverse adverse epigenetic regulation and modulate the gut microbiota, offering indirect yet potent strategies for cancer prevention [39]. Despite their potential, challenges persist, such as low solubility and undesirable side effects [45]. Moreover, their limited bioavailability has curtailed their clinical utility, necessitating exploration into alternative delivery methods like lipid-based systems to bolster efficacy [44]. There remains a pressing need for further research to refine and optimize phytochemical-based cancer treatments, striving toward enhanced efficiency and safety [45].

The journey toward developing safe and efficacious phytochemical-based treatments for cancer entails a nuanced and multifaceted process. Initially, the biological activity of plant materials is meticulously evaluated, whether in their dry or wet forms [42]. Subsequently, bioassay-guided fractionation is employed to pinpoint the active phytochemicals amidst the complex matrix of compounds (**Figure 3**) [53, 54]. Rigorous *in vitro* and *in vivo* assessments are conducted to determine the potency and effectiveness of the identified phytochemicals [55]. Encouragingly, phytochemicals have shown considerable promise in cancer therapy. Notable examples include diferuloylmethane from turmeric, epigallocatechin gallate (EGCG) from green tea extracts, resveratrol, and compounds derived from European mistletoe (*Viscum album*), with clinical evidence substantiating their anticancer effects [56]. Nevertheless, the intricacies of whole animal physiology, the tumor microenvironment, and the bioavailability of these phytochemicals during preclinical and clinical trials pose significant challenges that warrant attention [41]. These include poor bioavailability, safety concerns, low solubility, hydrophobicity limit, and a lack of pharmacodynamic reproducibility in clinical trials [57, 58]. Additionally, preclinical studies often use irrelevant or unachievable concentrations, highlighting the importance of considering clinical pharmacokinetics in experimental design [59]. Researchers are investigating novel delivery systems like nanoparticles, cyclodextrins, and liposomes to overcome these obstacles and enhance bioavailability and target specificity [58, 60]. The pleiotropic nature of phytochemicals necessitates careful balancing of efficacy and toxicity [61]. Moreover, factors such as whole animal physiology and tumor microenvironment need consideration [41]. Developing pharmacodynamic biomarkers is crucial for the

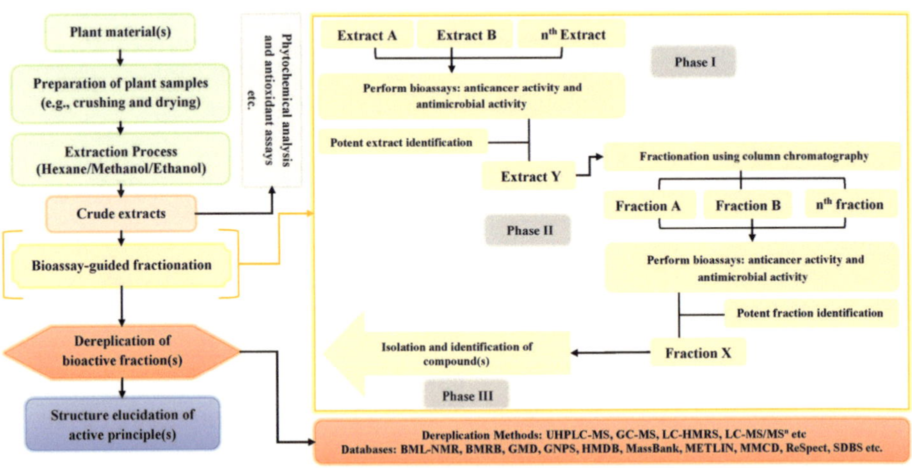


**Figure 3.**
*Schematic representation showing the various steps and techniques required during the characterization of plant bioactives including bioassay-guided fractionation of plant extracts.*

early assessment of biological activity and dose optimization in human trials [59]. The development of retinoids for cancer therapy and prevention exemplifies both the challenges and potential successes in this field [62]. An interdisciplinary approach and a thorough understanding of mechanisms of action are essential for improving the therapeutic translation of promising phytochemicals [57]. While complementary approaches cannot replace conventional cancer-destructive therapies, they may optimize treatment strategies when used alongside standard protocols [63].

Despite these obstacles, the burgeoning recognition of phytochemicals as a potential reservoir of effective and safer anticancer agents underscores their growing significance in the oncological landscape [64]. Overall, the rapidly expanding field of phytochemical research holds immense promise, offering natural compounds with chemopreventive and therapeutic potential against cancer. Further research efforts and controlled clinical studies are necessary to elucidate their mechanisms of action, optimize their delivery, and validate their clinical efficacy, paving the way for their integration into comprehensive cancer prevention, treatment strategies, and management [57].

## 4. Impact of phytochemicals on cancer treatment

The primary chemopreventive actions of phytochemicals involve reducing cancer risk by inhibiting the early phases of carcinogenesis, such as initiation and promotion, and enhancing the sensitivity of cancer cells to chemotherapy, potentially preventing metastasis. These naturally occurring compounds, abundant in fruits, vegetables, herbs, and spices, have garnered significant attention due to their diverse biological activities and potential therapeutic applications against various types of cancer [65–68].

### 4.1 Epidemiological perspective

Most epidemiological studies report significant benefits from increased phytochemical consumption and serum levels, linking higher exposure to these compounds with reduced cancer risk across various cancer types [69, 70]. Multiple systematic reviews and meta-analyses have identified associations between specific cancer types and dietary patterns determined through empirical analyses and predefined dietary indexes [71]. However, epidemiological studies do not always align with experimental findings. Translating these findings into clinical settings has proven challenging. Factors such as non-physiologically relevant concentrations, poor bioavailability, and the rapid metabolism of phytochemicals *in vivo* hinder their efficacy or different mechanisms of action than those observed in experimental settings [72]. Additionally, the complexity of exposure conditions and individual host factors contribute to the discrepancy between preclinical and clinical outcomes. Moreover, concerns about the use of phytochemicals include inconclusive evidence regarding their cancer prevention effects, inadequate information on potential side effects, insufficient assessment of pharmacodynamic properties, and conflicting findings regarding their molecular pathways, among other issues [73]. Furthermore, epidemiological research consistently suggests that fruit and vegetable consumption provides protective effects against cancer. However, due to the vast array of phytochemicals present in plants, pinpointing which specific compounds are responsible for these anticancer properties remains challenging [74]. Advancements in statistical methods and metabolomics, which focus on identifying objective biomarkers of dietary patterns, offer promising avenues for enhancing our understanding of how dietary factors influence cancer risk. These developments could elucidate the mechanisms underlying the disease-specific associations observed in epidemiological studies, despite the constraints of current animal models [71]. Despite the strong anticancer effects of phytochemicals and their proven efficacy in multiple epidemiological studies, many clinical trials have failed to replicate these findings [69]. The lack of evidence or the risk of harm has even led to the early withdrawal of some trials. Therefore, there is a pressing need for more human studies and clinical trials, with careful attention to safety measures, to fully understand the role of phytochemicals in cancer prevention and treatment [69].

### 4.2 Phytochemical modulation of tumor microenvironment and immunity

Organs, tissues, and cells collaboratively build the immune system, which protects the body from illnesses caused by microbes and abnormal cell transformations [75]. This system is regulated by macromolecules such as proteins, lipids, and carbohydrates [76–78]. Disruptions can lead to cancer, autoimmune diseases, allergies, and immunodeficiency disorders. The immune system maintains homeostasis by neutralizing harmful pathogens, toxins, and allergens [79]. The tumor microenvironment is pivotal in cancer progression, with immune cells playing a significant role [80, 81]. Intact immune responses can inhibit tumor development, whereas chronic inflammation can promote cancer growth [81]. Immunosuppressive cells, such as cancer-associated fibroblasts, tumor-associated macrophages, regulatory T-cells, and myeloid-derived suppressor cells, infiltrate tumors and impair cytotoxic T lymphocyte-mediated responses [82]. Additionally, metabolic competition between tumor cells and immune cells contributes to immunosuppression, as both require glucose and amino acids. Tumor cells alter their metabolism to produce immunosuppressive metabolites like lactate, which inhibit immune cell activity. These complex interactions between cancer cells and immune cells in the tumor microenvironment ultimately foster tumor growth and metastasis [83]. Targeting immunosuppressive cells and tumor metabolism presents promising avenues for cancer therapy [84]. Recent research underscores the potential of phytochemicals in cancer prevention and treatment through their immunomodulatory and antiangiogenic effects. Studies have demonstrated that phytochemicals can modulate the immune system in cancer patients by altering the quantity or function of immune cells in the tumor microenvironment, as evidenced by both *in vitro* and *in vivo* investigations [85–88]. These bioactive compounds enhance immune surveillance by stimulating T-cells, natural killer cells, and cytokine production, thereby boosting the body's ability to detect and destroy tumor cells. These substances can regulate various aspects of the immune response, including macrophage migration, T-cell stimulation, cytokine production, and immune cell reprogramming [86, 89]. For instance, phytochemicals like shikonin and hypericin induce immunogenic cell death, thereby enhancing tumor recognition by the immune system [90]. Additionally, these compounds exhibit antiangiogenic properties, inhibiting the formation of new blood vessels necessary for tumor growth and metastasis [91]. Moreover, they can modulate immune checkpoint molecules within the tumor microenvironment, potentially boosting the efficacy of immune checkpoint inhibitor therapy [88, 92]. Compounds such as curcumin, EGCG, and resveratrol have demonstrated the ability to target key molecular pathways within the tumor microenvironment, impacting various cell types and signaling cascades crucial for tumor progression [87]. Their ability to modulate diverse biological processes, coupled with their potential to mitigate the inflammatory microenvironment associated with cancer, positions phytochemicals as promising candidates for anticancer therapy [93].

### 4.3 Phytochemical synergy in cancer therapy

Phytochemicals show significant promise as adjunctive therapies in cancer therapy, enhancing the efficacy of conventional treatments while mitigating adverse effects [94, 95]. These compounds can enhance the efficacy of conventional chemotherapy and radiotherapy while mitigating their side effects [96, 97]. Phytochemicals exhibit a range of beneficial properties, including antioxidant, anti-inflammatory, and immunomodulatory effects, contributing to their protective role against treatment-induced toxicities [96]. Phytochemicals such as resveratrol and sulforaphane can potentiate the anti-tumor effects of chemotherapy and radiotherapy by sensitizing cancer cells and reversing chemoresistance [95]. Clinical studies have highlighted the potential of various phytochemicals, such as curcumin, ginsenosides, cannabinoids, epigallocatechin gallate, and resveratrol, in directly combating different types of cancer and alleviating chemotherapy-induced complications and toxicities [95]. Additionally, phytochemicals like quercetin, isoflavones, and catechins have shown potential for safeguarding non-cancerous cells from chemotherapy-induced damage [98]. While these findings are promising, further clinical trials are essential to fully assess the therapeutic potential of many phytochemicals in cancer management.

## 5. Mechanisms of plant-derived phytochemicals in cancer therapy

The molecular mechanisms underlying the anticancer effects of phytochemicals represent a fascinating and rapidly evolving area of research, shedding light on the intricate interactions between plant-derived compounds and cancer cells. Plant-derived phytochemicals exhibit diverse mechanisms of action that contribute to their potential in cancer prevention and treatment (**Figure 4**) [99]. Oncological research has demonstrated that phytochemicals and plant-based functional foods can induce apoptosis and suppress tumor cell survival, proliferation, invasion, and angiogenesis [100, 101]. They can inhibit the metabolic activation of pro-carcinogens or enhance the detoxification of carcinogens. Additionally, phytochemicals influence key cellular processes, including cell signaling, cell cycle regulation, oxidative stress, and inflammation. By mimicking the chemical structure and activity of various carcinogenic hormones, they disrupt harmful pathways. Furthermore, phytochemicals significantly impact epigenetic processes, such as DNA methylation, histone modification, and the regulation of non-coding RNAs. These actions help normalize altered cellular signaling in cancer cells. Notably, phytochemicals can modulate non-coding RNAs by increasing the expression of tumor-suppressing microRNAs (miRNAs) and decreasing the expression of oncogenic miRNAs (**Figure 5**) [102]. This regulation inhibits cancer cell growth and prevents cancer stem cells (CSCs) from self-renewing, thereby reducing the potential for cancer progression and recurrence [40].

### 5.1 Chronic inflammation and oxidative stress

Chronic inflammation is closely associated with cancer as it can lead to DNA damage, enhanced cell proliferation, and cancer progression [103]. The nuclear

**Figure 4.**
*Schematic representation of the key mechanisms through which phytochemicals impact and regulate various signaling pathways in cancer development and progression.*

factor κB (NF-κB) plays a key role as a regulator of the inflammatory response, and it is a link between persistent inflammation, cell proliferation, and tumor development (**Figure 6**) [104]. NF-κB is activated by external stimuli through canonical pathways involving mitogen-activated protein kinases (MAPKs), activator protein 1 (AP-1), pathogen-associated molecules (e.g., peptidoglycans), antigen receptors, and tumor necrosis factor-alpha (TNF-α). Additionally, specific TNF family proteins like lymphotoxin B and B-cell activating factor (BAFF) activate non-canonical pathways [105]. When activated, NF-κB moves to the nucleus where it controls the expression of over 100 genes, many of which are crucial for normal immune function [106]. However, dysregulation of NF-κB can lead to its overexpression, promoting cancer growth. The production of cyclooxygenase-2 (COX-2), an enzyme produced in response to cellular injury by epithelial cells, macrophages, and stromal cells, contributes to this process. COX-2 induction promotes cell proliferation and tumorigenesis by regulating key proteins involved in angiogenesis, immune evasion, and cell growth. Overproduction of prostaglandin E2 (PGE2) by COX-2 initiates inflammatory cascades that accelerate cell proliferation, inhibit apoptosis, and modulate immune responses, leading to premalignant conditions. PGE2 activates G-protein-coupled receptors (GPCRs), triggering intracellular anti-apoptotic pathways and promoting drug resistance in tumor cells [107].

In the tumor microenvironment, the excessive production of inducible nitric oxide synthase (iNOS) leads to inflammatory responses, which in turn trigger molecules like matrix metalloproteinases (MMPs) and vascular endothelial growth factor (VEGF). This series of events promotes tumor growth, angiogenesis, and metastasis.

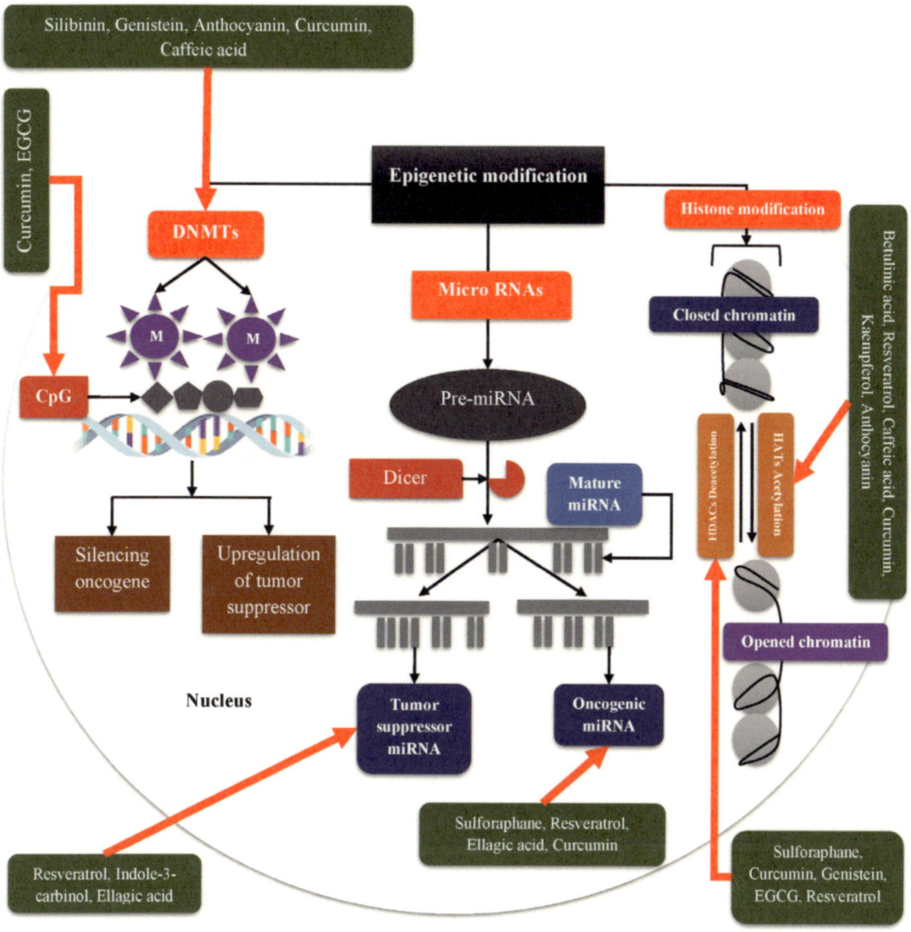


**Figure 5.**
*Epigenetic modifications induced by phytochemicals targeting DNA methylation, histone modifications, and miRNA-mediated gene regulation. This figure illustrates how phytochemicals influence these key epigenetic mechanisms to potentially reverse aberrant gene expression patterns associated with cancer.*

The nuclear factor kappa B (NF-κB) significantly influences the expression of both cyclooxygenase-2 (COX-2) and iNOS, either directly or indirectly. The interaction of multiple signaling pathways activates these molecules, contributing to the development of precancerous lesions and the progression of tumors [108]. Targeting COX-2, inducible nitric oxide synthase (iNOS), and NF-κB with selective inhibitors offers potential for cancer prevention. Several studies have shown that phytochemicals can interfere with the activation of these pathways, modulating inflammation and reducing cancer risk.

Furthermore, phytoconstituents play a crucial role in preventing cancer by creating antioxidant environments within normal cells, effectively countering the damaging effects of oxidative stress [109]. Elevated levels of reactive oxygen species (ROS) and a decline in the body's antioxidative defenses can lead to oxidative stress, disrupting physiological pathways and accelerating cancer development. Plant-based foods that are rich in polyphenols possess strong antioxidant properties, offering protection against oxidative stress-related diseases such as neurodegenerative disorders,

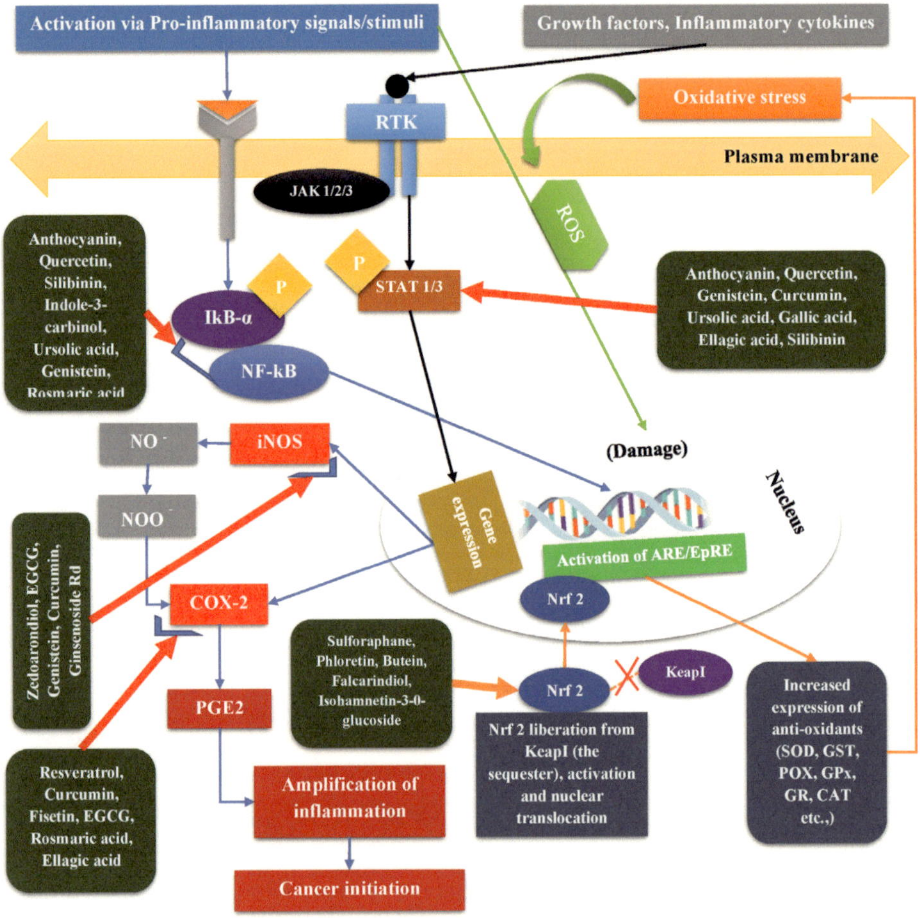


**Figure 6.**
*Mechanisms by which phytochemicals inhibit reactive oxygen species (ROS) and modulate STAT1-STAT3 signaling pathways, including the inflammatory responses mediated by COX-2, NF-κB, and iNOS. Activation of ARE/EpRE pathway by phytochemicals to prevent oxidative stress-induced cellular damage. ROS: reactive oxygen species; ARE: antioxidant response element; GST: glutathione-S-transferase; SOD: superoxide dismutase; Nrf2: nuclear factor erythroid 2 p45-related factor 2; EpRE: electrophile response element; Keap1: Kelch-like ECH-associated protein 1; p-signal transducer and activator of transcription (STAT3); NF-κB: nuclear factor kappa B; $NO^-$: nitric oxide radical; $NOO^-$: peroxynitrite radical; COX-2: cyclooxygenase 2; and iNOS: inducible nitric oxide synthase. Thick red arrows indicate suppression or inhibition of the corresponding pathways.*

atherosclerosis, diabetes, and cancer. The electron transfer capability of bioactive compounds underpins their antioxidative potential, enabling them to directly or indirectly counteract the harmful effects of ROS [110]. Many plant products contain phytochemicals that induce detoxifying enzymes, particularly in the liver, further enhancing the body's defense mechanisms [111]. These phytochemicals interact with the antioxidant response element (ARE) and the electrophile response element (EpRE), changing important molecular pathways. This pathway relies on the interplay between three key components: ARE, nuclear factor erythroid 2-related factor 2 (Nrf2), and Keap1 (Kelch-like ECH-associated protein 1). Nrf2 is released from its inhibitor Keap1 under oxidative stress, translocates to the nucleus, and activates the production of cytoprotective antioxidant enzymes [112, 113]. Dietary antioxidants can modulate ARE/EpRE signaling, significantly reducing the risk of cellular damage that contributes to carcinogenesis. By targeting these pathways, phytochemicals offer a powerful mechanism for reducing cancer risk and promoting long-term cellular health.

### 5.2 Apoptosis

The exploration of anticancer phytochemicals and their impact on apoptosis is a crucial aspect of cancer research, underscoring the potential of natural compounds to induce programmed cell death in cancer cells. Apoptosis, the process of programmed cell death, is a highly regulated mechanism essential for removing damaged or surplus cells. Researchers have identified phytochemicals as significant modulators of apoptosis, providing promising therapeutic avenues for cancer treatment [42].

These phytochemicals influence apoptosis through a variety of mechanisms, including activating intrinsic or extrinsic apoptotic pathways. For example, curcumin and resveratrol are known to activate the intrinsic apoptotic pathway by inducing mitochondrial dysfunction and the subsequent release of pro-apoptotic factors such as Cytochrome c. In contrast, compounds like epigallocatechin gallate (EGCG) from green tea can initiate the extrinsic pathway by upregulating death receptors and their ligands, leading to the activation of caspases and cell death. Additionally, phytochemicals often modulate key apoptotic regulators, including both pro-apoptotic and anti-apoptotic proteins [114]. As an example, quercetin raises the levels of the pro-apoptotic protein Bax while lowering the levels of the anti-apoptotic protein Bcl-2 (**Figure 7**). This causes cancer cells to undergo apoptosis. Similarly, studies have demonstrated that resveratrol enhances the activity of p53, a crucial tumor suppressor protein that promotes the expression of pro-apoptotic genes. These findings highlight the potential of phytochemicals in modulating apoptotic pathways to target cancer cells more effectively.

Under normal conditions, cells undergo apoptosis after prolonged cell cycle arrest, a process tightly regulated by a balance between pro-apoptotic and anti-apoptotic proteins [52]. However, cancer cells often disrupt this balance, resulting in uncontrolled cell division and the initiation of cancer. This disruption is frequently associated with mutations in tumor suppressor genes (TSGs), which enable cells to evade apoptosis and continue proliferating uncontrollably. Emerging evidence suggests that chemopreventive phytocompounds have the potential to restore apoptosis by targeting these disrupted pathways [115]. These phytochemicals may intervene in the apoptotic process, offering a promising strategy for cancer prevention and treatment (**Figure 7**).

### 5.3 Cell cycle and autophagy

Phytochemicals have gained significant attention for their potential anticancer properties, particularly through their impact on the cell cycle. A series of checkpoints and protein interactions meticulously regulate the cell cycle, which is essential for cell division and proliferation. Disruption in these regulatory mechanisms often results in uncontrolled cell growth, a defining characteristic of cancer. Phytochemicals can affect key regulators of the cell cycle, such as Cyclin-Cdk (Cyclin-dependent kinase) complexes. For instance, studies have shown that compounds like resveratrol and

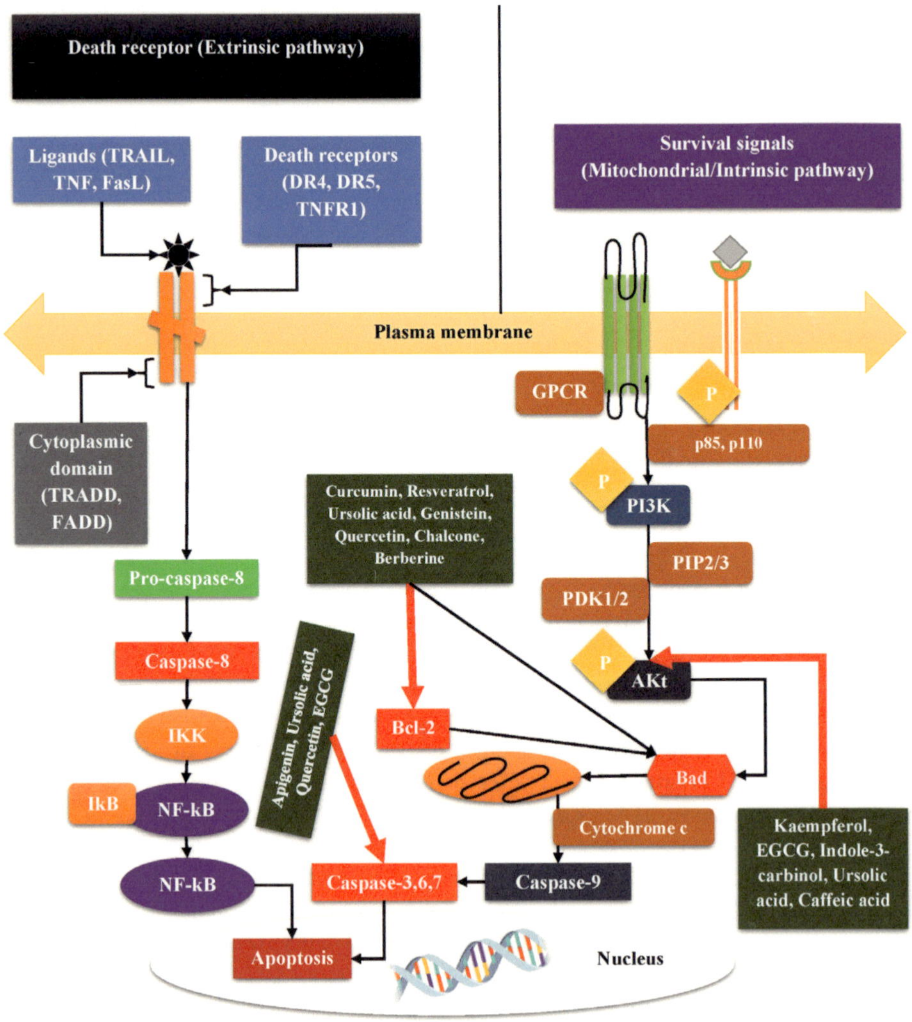


**Figure 7.**
*Effects of phytochemicals on apoptotic signaling pathways. Pro-apoptotic proteins include caspases, Bad, and IκB, while anti-apoptotic proteins include Bcl-2, PI3K, Akt, IKK, and NF-κB. Phytochemicals such as EGCG, resveratrol, isothiocyanates, Indole-3-carbinol, genistein, quercetin, and curcumin can inhibit anti-apoptotic proteins like Bcl-2 and stimulate pro-apoptotic proteins such as Bad. Specifically, EGCG, isothiocyanates, quercetin, and apigenin enhance caspase activity. Thick red arrows indicate suppression or inhibition of the corresponding pathways. TNF: tumor necrosis factor; FASL: FS-7-associated surface antigen ligand; TRAIL: targeting TNF-related apoptosis-inducing ligand; TNFR1: tumor necrosis factor receptor 1; DRS: death receptors; FADD: FAS-associated death domain protein; TRADD: TNFR1-associated death domain protein; and BCL2-associated agonist of cell death (BAD).*

genistein inhibit Cyclin-D1 and Cyclin-B1, disrupting the transitions from the G1 to S phase and the G2 to M phase. This disruption leads to cell cycle arrest and suppression of tumor growth (**Figure 8**).

Furthermore, phytochemicals can impact tumor suppressor proteins, such as p53 and Rb (retinoblastoma). Compounds like curcumin and apigenin enhance the activity of p53, which in turn increases the expression of its target proteins, including p21 and p27 [116]. These proteins inhibit Cyclin-Cdk complexes, leading to cell cycle

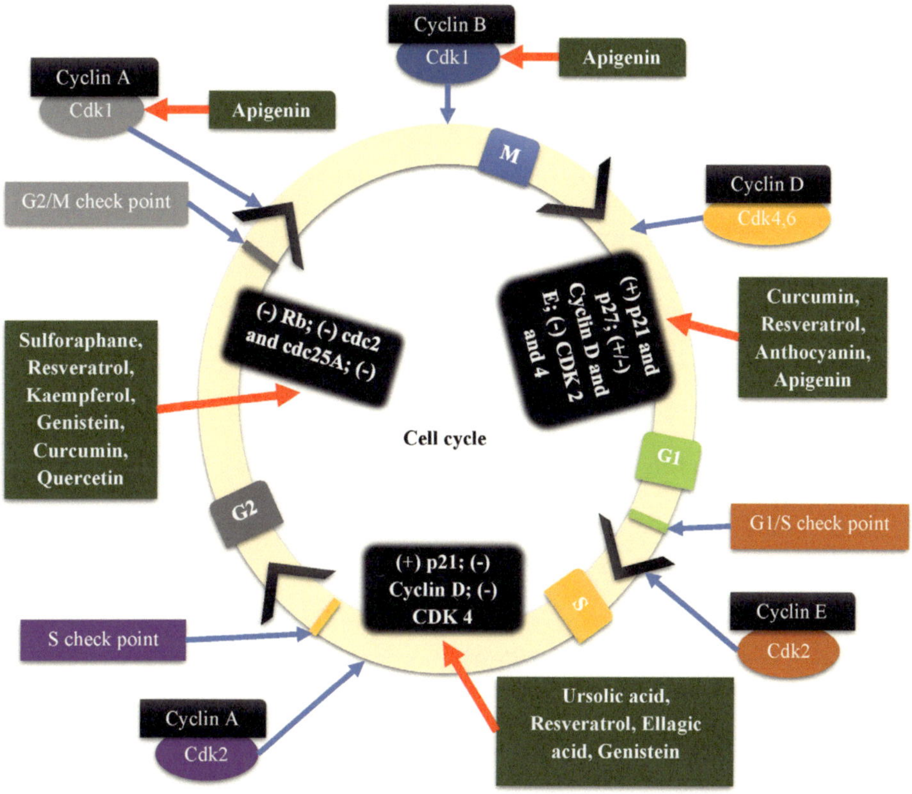


**Figure 8.**
*Effect of various anticancer phytochemicals on the cell cycle. The figure illustrates cell cycle phases with stimulatory proteins (+) and inhibitory proteins (−). The tumor suppressor protein p53 enhances the expression of p21 and p27, which inhibit Cyclin-Cdk complexes. Phytochemicals such as lycopene, quercetin, resveratrol, genistein, berberine, chalcone, and catechins arrest cells at the G2/M phase, while resveratrol and genistein also impact the S phase. Apigenin inhibits Cyclin A and B-Cdk1 complexes and promotes p21 and p27 expression, whereas resveratrol, curcumin, and apigenin enhance p53 activity. Thick red arrows indicate suppression or inhibition effect.*

arrest. Additionally, phytochemicals can modulate the phosphorylation status of critical proteins. For instance, lycopene and quercetin may influence the phosphorylation of Rb, affecting its ability to release the E2F transcription factor and thereby modulating cell cycle progression. At specific phases, phytochemicals also induce cell cycle arrest. Lycopene, quercetin, resveratrol, and genistein are known to arrest cells in the G2/M phase, while resveratrol impacts the S phase [117, 118]. This selective targeting enables the development of tailored therapeutic strategies based on the cell cycle phase most critical for tumor suppression.

Autophagy, a cellular process involving the degradation and recycling of cellular components, plays a dual role in cancer. It can either suppress tumor initiation by removing damaged organelles and proteins, or it can promote cancer cell survival under stress. Researchers have shown that phytochemicals like curcumin, resveratrol, and quercetin influence autophagy through various mechanisms [119, 120]. Curcumin, for example, induces autophagic cell death by upregulating key autophagy-related proteins like LC3 and Beclin-1 while downregulating p62, a protein associated with autophagy inhibition [121]. Resveratrol activates autophagy through the AMPK pathway and inhibits mTOR signaling, leading to enhanced cancer cell degradation. Quercetin, on the other hand, modulates autophagy by affecting the PI3K/Akt/mTOR pathway, contributing to its anticancer effects.

Phytochemicals can also influence the balance between autophagy and apoptosis in cancer cells [114]. For example, certain compounds may induce autophagic cell death in cancer cells, providing a therapeutic advantage by overcoming resistance to conventional treatments. Additionally, phytochemicals like epigallocatechin gallate (EGCG) from green tea and sulforaphane from broccoli have shown promise in promoting autophagic cell death, which could enhance their efficacy as part of integrated cancer therapy. Overall, anticancer phytochemicals modulating autophagy represent a promising avenue for developing novel cancer treatments, potentially improving therapeutic outcomes by leveraging the dual role of autophagy in cancer biology [122].

### 5.4 Pro-survival signaling (Wnt/β-catenin; PI3K/Akt/mTOR; JAK/STAT pathways)

#### *5.4.1 Wnt/β-catenin signaling pathway*

The Wnt/β-catenin signaling pathway is pivotal in the self-renewal and maintenance of cancer stem cells (CSCs), making it a critical target in cancer progression [40]. Central to this pathway is the Wnt ligand, a modified glycoprotein secreted into the extracellular matrix. In the absence of Wnt signaling, β-catenin is phosphorylated within a destruction complex comprising glycogen synthase kinase 3β (GSK3β), casein kinase 1α (CK1α), AXIN2, and adenomatous polyposis coli (APC) protein. This phosphorylation marks β-catenin for ubiquitination by β-transducin repeat-containing protein (β-TRCP), leading to its subsequent degradation by the proteasome. Upon Wnt activation, the Wnt ligands bind to Frizzled (FZD) and lipoprotein receptor-related protein (LRP)-5/-6 receptors, inhibiting the degradation of β-catenin. This binding triggers the phosphorylation of LRP5/6 by CK1α and GSK3β, facilitating the translocation of AXIN2 to the membrane, where it interacts with phosphorylated LRP5/6 and Disheveled (DVL) proteins (**Figure 9**). The recruitment of DVL disrupts the degradation complex, allowing β-catenin to accumulate in the cytoplasm. Stabilized β-catenin then translocates into the nucleus, forming a complex with lymphoid enhancer factor/T-cell factor (LEF/TCF) and various co-activators, driving the transcription of target genes. These genes, including Fos-related antigen 1, Myc, Cyclin D1, ABC transporters, MMP7, CD44, CXCL12, CXCR4, Fibronectin, and COX2, are integral to regulating cell polarity, cytoskeletal dynamics, and differentiation, all of which contribute to cancer progression and metastasis.

Anticancer phytochemicals have demonstrated significant potential for modulating the Wnt/β-catenin signaling pathway, a critical regulator of cellular processes such as proliferation, differentiation, and stem cell renewal [100]. In cancer, aberrant activation of this pathway contributes to tumorigenesis and progression, particularly through the maintenance of cancer stem cells (CSCs) and the promotion of epithelial-to-mesenchymal transition (EMT). Phytochemicals such as curcumin, resveratrol, quercetin, and epigallocatechin gallate (EGCG) have been shown to inhibit the Wnt/β-catenin signaling cascade, leading to reduced β-catenin levels in the cytoplasm and nucleus (**Figure 9**). For instance, curcumin can downregulate Wnt ligands and receptors, thereby preventing the accumulation of β-catenin and its transcriptional activity in the nucleus. Resveratrol has been found to enhance the degradation of

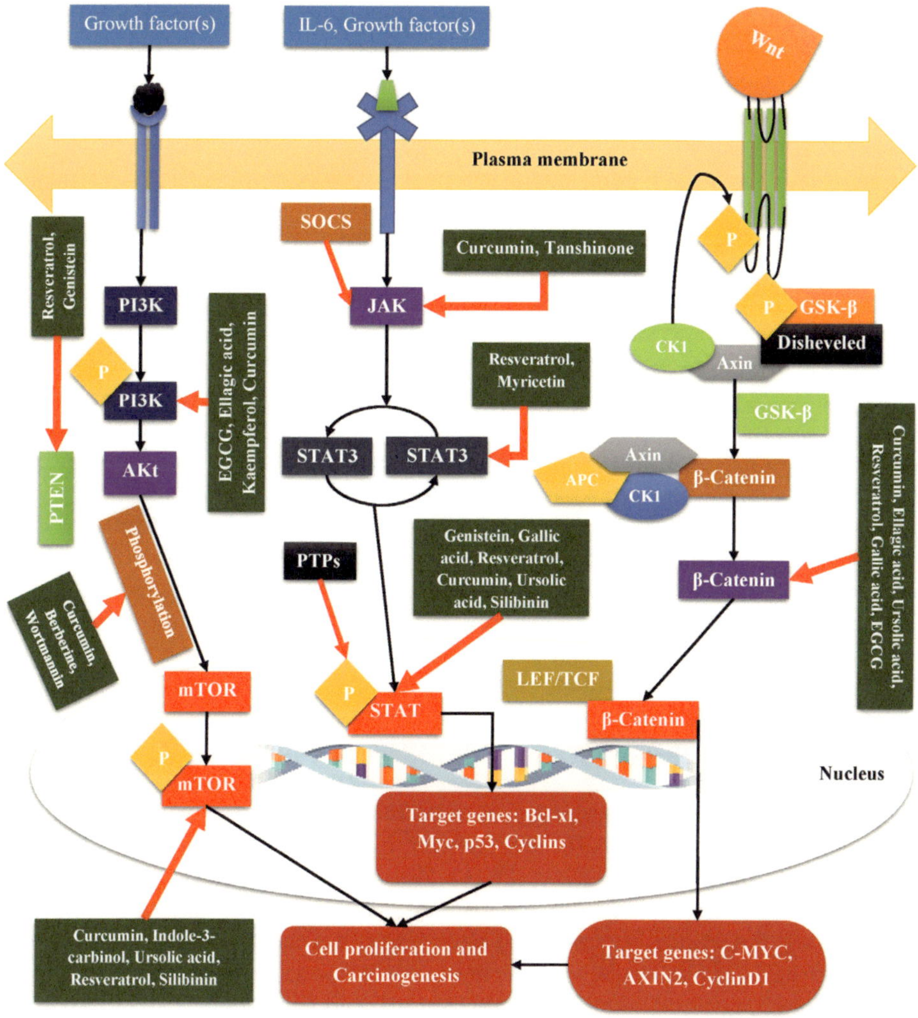


**Figure 9.**
*Illustration of the Wnt/β-catenin, PI3K/Akt/mTOR, and JAK/STAT signaling pathways and their roles in cancer progression. This figure highlights how various phytochemicals target specific components within these pathways to inhibit cancer progression. Key components and interactions are depicted: Wnt/β-catenin pathway (involving Wnt ligands, β-catenin, Frizzled receptors, and LEF/TCF complexes); PI3K/Akt/mTOR pathway (featuring PI3K, Akt, mTOR, PTEN, and cyclin-dependent kinases (CDKs)); JAK/STAT pathway (showing IL-6, JAK, STAT, SOCS, and anti-apoptotic proteins like Mcl-2 and Bcl-xl). The effect of phytochemicals on these pathways is illustrated, indicating their potential to inhibit various steps in these cascades. Abbreviations: PI3K: phosphatidylinositol-3-kinase; PTEN: phosphatase and tensin homolog; mTOR: mammalian target of rapamycin; CDKs: cyclin-dependent kinases; IL-6: interleukin 6; JAK: Janus kinase; STAT: signal transducer and activator of transcription; Mcl-2: myeloid cell leukemia 2 protein; SOCS: suppressor of cytokine signaling; Bcl-xl: B cell lymphoma extra-large; PTPs: protein tyrosine phosphatases.*

β-catenin via increased activity of the destruction complex, while quercetin inhibits the transcriptional activity of β-catenin target genes such as Cyclin D1 and Myc. EGCG similarly interferes with β-catenin translocation and activation, suppressing the downstream transcription of genes involved in cancer cell proliferation and survival. By disrupting the Wnt/β-catenin axis, these phytochemicals can impede cancer cell growth, promote apoptosis, and reduce CSC populations, presenting a promising therapeutic avenue for targeting cancers driven by aberrant Wnt signaling.

### *5.4.2 PI3K/Akt/mTOR and JAK/STAT signaling pathways*

Several critical signaling pathways that regulate cell survival in response to DNA damage are integral to cancer development and spread. Key pathways of interest include the Janus kinase (JAK)/signal transducer and activator of transcription (STAT) pathway and the phosphoinositide 3-kinase (PI3K)/Akt/mammalian target of rapamycin (mTOR) pathway [101].

The PI3K/Akt/mTOR signaling pathway plays a pivotal role in regulating cell growth, proliferation, survival, and metabolism [123]. The PI3K/Akt/mTOR signaling cascade is often disrupted in cancer, particularly in liver cancers, where it promotes tumor growth and survival [41]. Aberrant activation of this pathway leads to upregulation of anti-apoptotic factors such as NF-κB and cyclin D1, contributing to carcinogenesis. Akt activation and subsequent phosphorylation in response to inflammatory signals further enhance the cancerous process by targeting nuclear factors like mTOR. Inhibiting the PI3K/Akt/mTOR pathway presents a promising strategy for cancer therapy [124].

The JAK/STAT signaling pathway plays a crucial role in regulating immune responses, cell growth, and differentiation. Similarly, the JAK/STAT pathway plays a crucial role in inflammation, neoplastic transformation, tumor progression, metastasis, and angiogenesis [99]. Proteins such as suppressor of cytokine signaling (SOCS), protein tyrosine phosphatases (PTPs), and protein inhibitors of activated STATs (PIAS) regulate the pathway's activity by modulating both basal and induced expressions of JAKs and STATs. Targeting these regulatory proteins with plant-based chemotherapeutic compounds could represent a novel approach for cancer prevention and treatment.

Plant-derived phytochemicals have demonstrated the potential to inhibit cell proliferation by modulating these pathways. For instance, curcumin has been shown to inhibit PI3K, an upstream activator of the Akt pathway [125]. By reducing PI3K activity, curcumin effectively suppresses downstream signaling, thereby impeding tumor growth. Additionally, the mTOR kinase, which is crucial for cell growth and protein synthesis, is another key target for phytochemical intervention. Compounds such as indole-3-carbinol, ursolic acid, resveratrol, and silibinin have demonstrated the capacity to inhibit mTOR activity (**Figure 9**). Phytochemicals like curcumin and resveratrol have been shown to impact the JAK/STAT pathway. Curcumin inhibits JAK2 activity by disrupting STAT3 phosphorylation and its subsequent nuclear translocation. Similarly, resveratrol can inhibit the JAK2/STAT3 signaling axis, reducing the expression of genes involved in tumor growth and metastasis. Other phytochemicals, such as genistein, inhibit JAK2 kinase activity, leading to decreased STAT3 signaling and enhanced apoptosis in cancer cells. Overall, these phytochemicals offer a multifaceted approach to disrupting the PI3K/Akt/mTOR and JAK/STAT pathway, underscoring their potential as complementary or alternative therapies in cancer treatment.

### 5.5 Epigenetic modifications

Epigenetics examines heritable gene expression changes that do not alter the DNA sequence itself. Epigenetic mechanisms, including DNA methylation, histone modifications, and RNA-mediated gene silencing, significantly contribute to cellular differentiation and development, and they are closely associated with the onset and advancement of cancer (**Figure 5**) [125]. DNA methylation involves the addition of a methyl group to cytosine residues within CpG dinucleotides, resulting in the formation of 5-methylcytosine. The DNA methyltransferase (DNMT) family of enzymes catalyzes this modification: DNMT1 acts as a maintenance methyltransferase, and DNMT3A and DNMT3B act as *de novo* methyltransferases [126]. Although CpG dinucleotides are sparse throughout the genome, their hypermethylation in tumor suppressor gene (TSG) CpG islands can lead to the silencing of these genes and contribute to the inactivation of TSG functions in cancer cells. Histone modifications, specifically acetylation, and methylation of lysine-rich histone tails, are pivotal in regulating chromatin structure and gene expression. Histone methyltransferases (HMTs) regulate histone methylation, while histone modification acetyltransferases (HATs) and histone deacetylases (HDACs) control histone acetylation and deacetylation, respectively [127]. Disruptions in these histone modification patterns are associated with carcinogenesis, often resulting in the silencing of TSGs [126]. Furthermore, researchers are increasingly recognizing the roles of enzymes involved in these modifications in modulating the activity of both histone and non-histone proteins, thereby influencing cellular processes in both normal physiology and cancer pathogenesis [128, 129].

Anticancer phytochemicals exert significant effects on epigenetic modifications, which are crucial in regulating gene expression without altering the DNA sequence (**Figure 5**). Inverse epigenetic alteration is a target for chemopreventive and therapeutic purposes, with phytochemicals potentially reversing these changes to prevent colon cancer [125]. Studies show that phytochemicals like curcumin and resveratrol can inhibit DNA methyltransferases (DNMTs), which add methyl groups to DNA, reactivating silenced tumor suppressor genes and reducing cancer progression [130]. These compounds also affect histone modifications, such as acetylation and methylation, which are crucial for chromatin structure and gene expression. Resveratrol increases histone acetylation, enhancing the expression of tumor suppressor genes. EGCG inhibits histone deacetylases, leading to increased acetylation and activation of genes that suppress cancer growth. Some phytochemicals also impact RNA-mediated gene-silencing mechanisms, such as ellagic acid modulating microRNA expression [131]. Phytochemicals have the potential to modify epigenetic processes, offering a promising approach to preventing and treating cancer by reversing abnormal gene expression patterns and reactivating genes that suppress tumor growth and progression.

## 6. Challenges in current cancer treatments

Current cancer treatments face numerous challenges, largely due to the complex and heterogeneous nature of tumors (**Figure 10**) [132, 133]. Tumor heterogeneity arises from various genetic and epigenetic modifications, resulting in diverse subpopulations of cancer cells within a single tumor. The inherent plasticity of cancer cells, which enables them to undergo phenotypic switching at the genetic, metabolic, and mechanistic levels, further compounds this heterogeneity. Phenotypic switching plays a key role in cancer drug resistance, a major factor contributing to treatment failure. Conventional cancer therapies primarily target proliferative cancer cells but fail to eliminate a small yet critical subset of tumor-initiating cells known as cancer stem cells (CSCs) [40, 134]. These cells are characterized by their quiescence, drug resistance, immune evasion, epithelial-to-mesenchymal transition (EMT), mesenchymal-to-epithelial transition (MET), and metastatic potential. CSCs are not only resistant

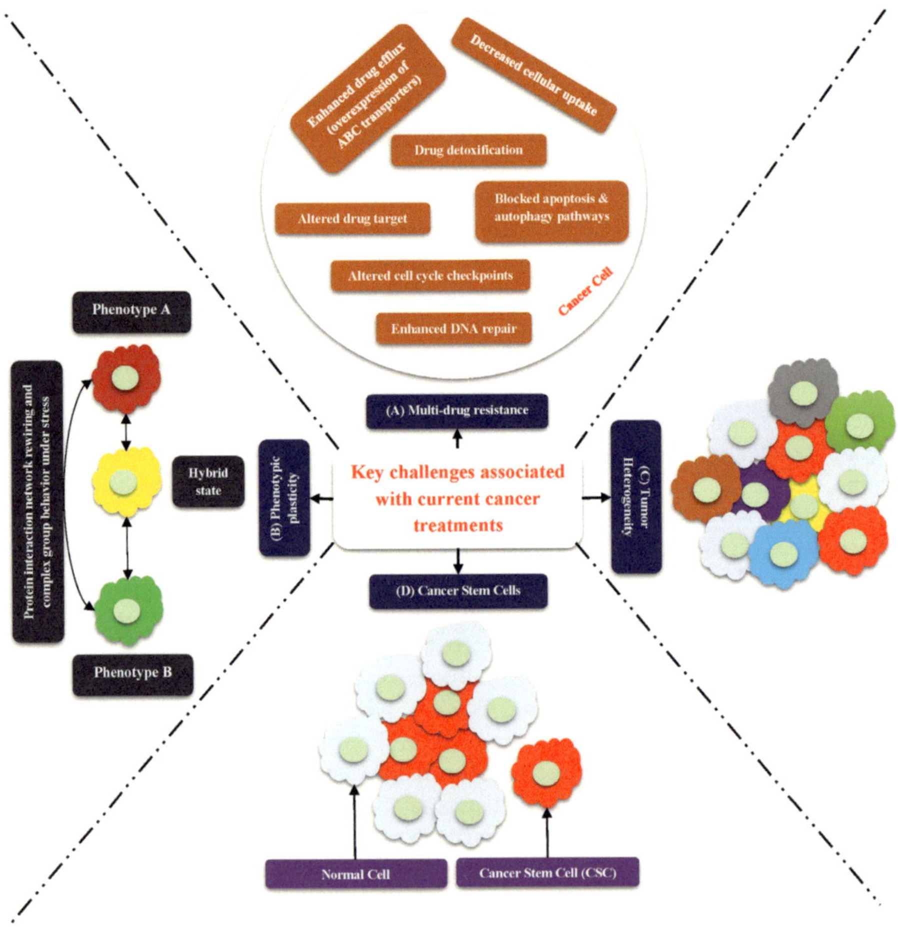


**Figure 10.**
*A schematic representation of key challenges in current cancer treatments. (A) Various factors contribute to multi-drug resistance (MDR), including drug efflux, altered drug targets, and enhanced DNA repair mechanisms. (B) Phenotypic plasticity in cancer cells, driven by the rewiring of protein interaction networks, plays a pivotal role in MDR, allowing dynamic adaptation and facilitating the transition from drug-sensitive to drug-resistant phenotypes. (C) Tumor heterogeneity arises from a range of genetic and epigenetic modifications, contributing to treatment resistance and variable patient outcomes. (D) Conventional cancer therapies primarily target actively dividing cancer cells, often overlooking cancer stem cells (CSCs), which possess distinct traits such as quiescence, drug resistance, immune evasion, epithelial-mesenchymal transition (EMT), and metastasis potential, leading to recurrence and treatment failure.*

to most current treatments but also serve as the driving force behind cancer relapse and metastasis. Their ability to evade conventional therapies and survive makes CSCs a pivotal obstacle in cancer treatment [40]. Therefore, effective targeting of CSCs is essential for achieving long-term therapeutic success, underscoring the urgent need for alternative and advanced cancer treatment strategies.

Over the past few decades, extensive research has highlighted the broad-spectrum anticancer potential of phytochemicals, which exhibit properties such as inhibition of tumor growth, induction of apoptosis, and modulation of key signaling pathways involved in cancer progression. However, their clinical application remains limited due to inherent challenges such as poor aqueous solubility, low bioavailability, and rapid clearance from the body, all of which reduce their therapeutic efficacy [135]. One promising strategy to overcome these limitations is the use of nanoparticle-based drug delivery systems, which passively target tumor sites through the enhanced permeability and retention (EPR) effect—a phenomenon that allows nanoparticles to accumulate in tumor tissue due to leaky vasculature and poor lymphatic drainage [136]. Nano-phytomedicine, an emerging field that integrates phytochemicals and nanotechnology, provides an innovative solution by improving the pharmacokinetic and pharmacodynamic profiles of these compounds. Encapsulating phytochemicals in nanoparticles enhances their stability, bioavailability, and retention in the body, potentially leading to more effective cancer therapies with reduced systemic toxicity and side effects [137]. This approach not only maximizes the therapeutic potential of phytochemicals but also opens new avenues for precision medicine in cancer treatment (**Figure 11**).

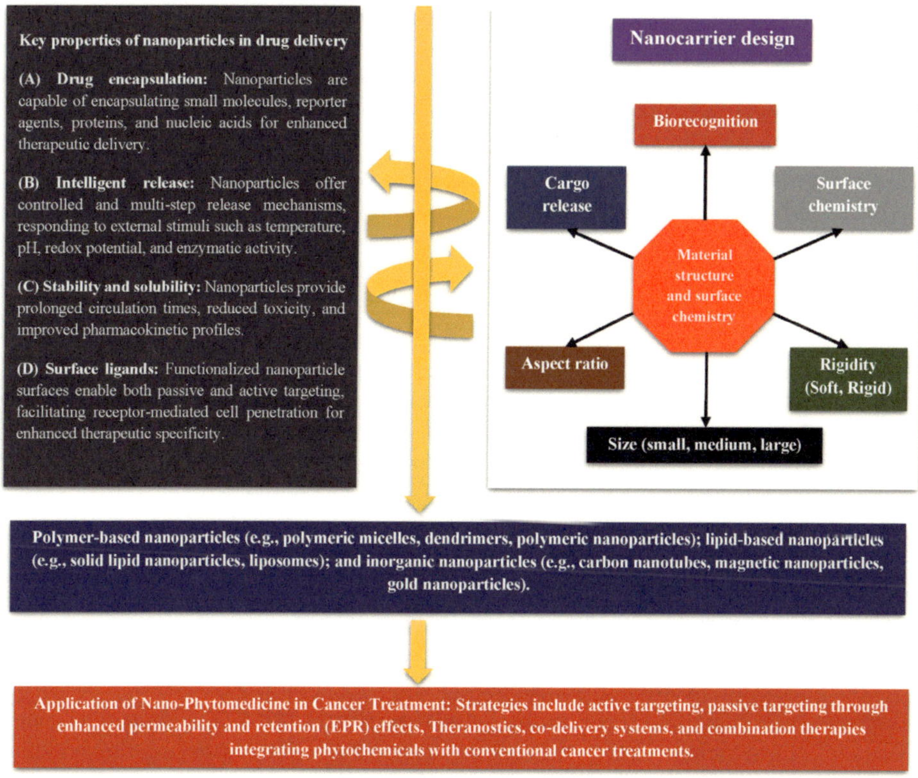


**Figure 11.**
*An overview of nanocarrier design and the applications of nano-phytomedicine in cancer treatment. The figure illustrates various strategies for enhancing therapeutic efficacy, including (1) active targeting—direct delivery of nano-phytomedicines to specific cancer cells through surface modifications with targeting ligands; (2) passive targeting via EPR Effects—utilization of the enhanced permeability and retention (EPR) effect to accumulate nanomedicines in tumor tissues; (3) theranostics—integration of therapeutic and diagnostic functions within a single nanoformulation for simultaneous treatment and imaging; (4) co-delivery—combination of phytochemicals with other therapeutic agents in a single nanocarrier to enhance therapeutic outcomes; and (5) combination therapy—synergistic use of phytochemicals with conventional cancer treatments (e.g., chemotherapy, radiotherapy) to improve overall treatment efficacy and reduce side effects.*

## 7. Nanoparticle-based phytomedicines in cancer therapy

Nanomedicine's advancement holds significant promise for revolutionizing cancer therapy by improving drug delivery specificity and efficacy (**Figure 11**) [138, 139]. The ultimate aim is to optimize phytochemical-nanocarrier systems to achieve targeted and efficient delivery of therapeutic agents to cancer cells while minimizing systemic toxicity [140]. This goal can be approached through several key strategies, including:

- *Designing phytochemical-nanocarriers*: the primary objective is to engineer nanocarriers that improve the delivery and uptake of phytochemicals, specifically within cancer cells. By encapsulating phytochemicals in nanoscale carriers, the goal is to boost their therapeutic effectiveness and minimize off-target [137]. This approach not only increases the concentration of the drug at the tumor site but also reduces adverse effects on healthy tissues, thereby improving the overall safety profile in clinical settings.

- *Enhancing tumor targeting*: to further refine the specificity of nano-phytomedicine, it is essential to conjugate nanocarriers with targeting ligands that bind selectively to tumor-associated receptors. Active targeting strategies utilize these ligands to precisely direct nanocarriers to cancer cells, facilitating improved accumulation and internalization of phytochemicals within the tumor microenvironment. This targeted approach enhances therapeutic outcomes and reduces the likelihood of drug resistance.

- *Achieving site-specific drug release*: one of the significant advantages of nanomedicine is its potential for site-specific drug release. By designing nanocarriers that release their payload in response to specific tumor microenvironmental triggers, such as pH or enzymatic activity, nonspecific drug distribution can be prevented and systemic toxicity minimized [136]. This controlled release mechanism ensures that the therapeutic agents exert their effects precisely where needed, thereby optimizing treatment efficacy and reducing side effects.

The development of nano-phytomedicine that leverages these strategies represents a promising frontier in cancer research. By focusing on targeted delivery, enhanced tumor localization, and controlled release, the clinical outcomes of phytochemical-based therapies can be improved, effectively addressing the challenges associated with conventional treatment approaches.

Taking a smart, realistic, and rational approach can enhance the clinical impact of nano-phytomedicine. Strategic directions are outlined for improving the performance, translation, and application of nano-phytomedicine in cancer treatment.

### 7.1 Enhancing tissue-specific targeting and half-life enhancement in nano-phytomedicine

The challenge of optimizing the tissue-specific targeting and extending the half-life of nano-phytomedicine in systemic administration is a significant concern in cancer therapy. Achieving targeted delivery of phytochemicals to specific tissues while minimizing off-target effects is crucial for maximizing therapeutic efficacy and reducing systemic toxicity. PEGylation, the process of covalently or non-covalently attaching polyethylene glycol (PEG) chains to therapeutic molecules or nanoparticles, offers a promising solution to address these challenges [141].

PEGylation could potentially confer nanoparticles with a "stealth" property, primarily due to its hydrophilic nature, which facilitates the formation of a hydration layer. This hydration layer enhances water solubility and decreases protein adsorption on their surface. This reduction in protein binding minimizes the formation of a protein corona, leading to decreased recognition and clearance by the immune system. Consequently, PEGylated nanoparticles exhibit prolonged circulation times in the bloodstream, enabling them to evade the reticuloendothelial system (RES) and significantly reduce their antigenicity. These advantageous properties have the potential to significantly improve the effectiveness of nanoparticles in drug delivery and targeted therapies, ultimately enhancing treatment outcomes while minimizing associated side effects. Therefore, PEGylation has the potential to effectively shield nanoparticles from detection and uptake by the immune system, particularly by macrophages, and enhance their biocompatibility. This stealth characteristic is crucial for improving the circulatory half-life of nano-phytomedicines, allowing them to remain in the bloodstream longer and thereby increasing the probability of reaching and accumulating in the targeted tissues or tumor sites.

By reducing immune recognition and clearance, PEGylated nanomedicines exhibit prolonged systemic circulation, which is essential for effective cancer therapy. This extended half-life improves the pharmacokinetics of phytochemicals, allowing for sustained therapeutic concentrations and enhanced therapeutic efficacy. Furthermore, PEGylation can be tailored to optimize the size, surface charge, and other physicochemical properties of nanoparticles, further improving their tissue-specific targeting and delivery.

Overall, incorporating PEGylation into the design of nano-phytomedicines represents a strategic advancement in enhancing their therapeutic potential, improving targeting precision, and extending their circulatory half-life, ultimately contributing to more effective and less toxic cancer treatments.

### 7.2 Combination of nano-phytomedicines with current immunotherapeutic approaches

Recent advancements in cancer immunotherapy have demonstrated substantial progress in treating various malignancies. However, current therapies often face significant pathophysiological constraints that can limit their efficacy and patient outcomes. These limitations include immune evasion by tumors, insufficient immune activation, and adverse effects associated with conventional treatments.

To address these challenges, recent research has focused on developing innovative approaches that combine traditional immunotherapies with advanced nanotechnology-based solutions. One promising avenue is the use of immunomodulatory nano-phytomedicines, which integrate phytochemicals with nanotechnology to enhance the immune response against cancer cells. These nano-phytomedicines have the potential to modulate immune pathways, increase immune targeting specificity, and reduce the systemic toxicity commonly associated with conventional therapies.

Additionally, the combination of existing immunotherapies with nano-phytomedicine or DNA insert-encapsulated nanoformulations, particularly for CAR-T (chimeric antigen receptor T-cell) therapy, offers a novel strategy to overcome current limitations. By incorporating nano-phytomedicines or nanoformulations into CAR-T therapy, it is possible to enhance the activation and persistence of engineered T-cells within the patient's body. This approach may improve the efficacy of CAR-T therapy by enhancing immune cell functionality and promoting better tumor targeting [132].

This combined approach includes key aspects:

- *Enhanced immune activation*: immunomodulatory nano-phytomedicines can boost the immune system's ability to recognize and attack cancer cells, potentially overcoming tumor-induced immune suppression.
- *Targeted delivery*: nanoformulations, including DNA inserts encapsulated in nanoparticles, can facilitate targeted delivery of therapeutic agents to specific tumor sites or immune cells, thereby increasing therapeutic precision and reducing off-target effects.
- *Improved CAR-T cell efficacy*: by combining nano-phytomedicines with CAR-T therapy, it is possible to enhance the persistence and effectiveness of engineered T-cells, leading to more durable responses and increased survival rates.
- *Reduction of adverse effects*: combining immunotherapies with nano-phytomedicines may help mitigate some of the side effects associated with conventional treatments, contributing to a better overall patient experience.

By advancing the integration of nano-phytomedicines with current immunotherapeutic approaches, this strategy holds the potential to significantly improve immuno-oncological outcomes and increase the number of long-term cancer survivors. This innovative approach represents a forward-looking perspective in cancer treatment, leveraging the strengths of both immunotherapy and nanotechnology to overcome existing challenges and enhance patient care.

## 7.3 Combination of conventional chemotherapies with phytochemicals

Numerous studies have highlighted a critical limitation of conventional cancer therapies: while these treatments are effective against rapidly dividing cancer cells, they often fail to eradicate quiescent cancer stem cells (CSCs) within the tumor mass. CSCs are characterized by their ability to self-renew, differentiate, and contribute to tumor heterogeneity, making them a significant challenge in achieving complete tumor regression [40, 132, 142]. This residual population of CSCs can lead to disease recurrence and metastasis, undermining the efficacy of traditional treatments.

In this context, recent research suggests that phytochemicals possess potent anti-CSC properties. Natural sources provide phytochemicals that specifically target and disrupt key pathways involved in CSC maintenance and function. These compounds can induce apoptosis, inhibit self-renewal, and affect the signaling pathways critical for CSC survival, such as the Wnt/β-catenin, Notch, and Hedgehog pathways [32, 39, 70, 99, 125].

Combining conventional chemotherapeutics with phytochemicals presents a promising strategy to address both cancer cells and CSCs, potentially leading to more effective and comprehensive tumor eradication [143]. This approach leverages the strengths of each treatment modality.

- *Enhanced efficacy against diverse tumor populations*: conventional therapies target the bulk of rapidly proliferating cancer cells, while phytochemicals can specifically target CSCs, addressing the resistant subpopulation often responsible for relapse and metastasis.

- *Drug resistance reduction*: phytochemicals can modify the tumor microenvironment and influence drug resistance mechanisms, potentially increasing CSCs' sensitivity to conventional treatments.

- *Synergistic effects*: the combined use of phytochemicals and conventional chemotherapeutics can result in synergistic effects, amplifying the overall therapeutic impact and reducing the likelihood of CSC-driven tumor relapse [42].

- *Improved clinical outcomes*: integrating phytochemicals into treatment regimens could enhance the overall therapeutic efficacy, leading to better clinical outcomes, including prolonged remission and a reduced likelihood of recurrence.

Therefore, incorporating phytochemicals into conventional cancer treatment regimens offers a strategic advantage by targeting both the proliferative cancer cells and the resilient CSCs. This combination approach holds the potential for achieving more effective tumor regression, overcoming treatment resistance, and ultimately improving patient outcomes.

### 7.4 Co-delivery of phytochemicals and RNA therapeutics through nanomedicine for enhanced cancer therapy

The co-delivery of phytochemicals with microRNA (miRNA) or small interfering RNA (siRNA) through nanomedicine represents a promising strategy to enhance cancer treatment by targeting multiple facets of tumor biology [139]. This integrated approach has the potential to significantly improve therapeutic outcomes by simultaneously modulating cancer cells, cancer stem cells (CSCs), and their surrounding microenvironment [140].

1. *Synergistic effects on cancer cells and CSCs*: combining phytochemicals with miRNA or siRNA in a single nanomedicine formulation can offer a multifaceted approach to cancer therapy. Phytochemicals, known for their ability to induce apoptosis, inhibit proliferation, and suppress angiogenesis, can be effectively delivered to cancer cells to exert direct cytotoxic effects. Concurrently, miRNAs or siRNAs can be used to silence specific genes involved in cancer progression or resistance mechanisms. For instance, miRNAs that target oncogenes or genes responsible for CSC maintenance can be co-delivered with phytochemicals to enhance the overall therapeutic impact.

2. *Tumor microenvironment modulation*: the tumor microenvironment (TME) plays a crucial role in cancer progression and resistance to therapy. Phytochemicals can influence the TME by reducing inflammation, inhibiting fibrosis, and modulating immune responses. This approach, when combined with RNA therapeutics that target TME-associated factors, such as those involved in immune evasion or extracellular matrix remodeling, can further enhance therapeutic efficacy. For example, siRNA targeting pro-inflammatory cytokines or angiogenic factors can be co-delivered with phytochemicals to simultaneously address both tumor cells and their supportive environment.

3. *Enhanced delivery and reduced toxicity*: nanomedicine enables the encapsulation and targeted delivery of both phytochemicals and RNA molecules. Nanoparticles can be engineered to enhance the stability, bioavailability, and specific delivery of these therapeutic agents to tumor sites. This targeted delivery reduces systemic toxicity and improves the therapeutic index. Additionally, by utilizing nanocarriers that can release both types of therapeutics in a controlled manner, it is possible to achieve a synergistic effect while minimizing adverse side effects.

4. *Overcoming resistance mechanisms*: cancer cells and CSCs often develop resistance to conventional therapies through a variety of mechanisms, including upregulation of survival pathways and downregulation of apoptosis-related factors. Co-delivering phytochemicals with miRNA or siRNA can address resistance mechanisms more effectively. For example, phytochemicals might sensitize cells to RNA-based therapies, while RNA therapeutics can modulate pathways that confer resistance, leading to improved treatment responses.

5. *Personalized therapeutic approaches*: individual patient profiles, based on specific genetic and molecular characteristics of their tumors, can tailor the combination of phytochemicals and RNA-based therapies. This customized approach improves treatment precision by tailoring both the chemical and RNA components to the patient's cancer biology.

As such, the co-delivery of phytochemicals with miRNA or siRNA via nanomedicine offers a sophisticated and multi-pronged approach to cancer treatment. By targeting cancer cells, CSCs, and the TME simultaneously, this strategy holds promise for improving therapeutic outcomes, overcoming resistance, and advancing personalized medicine in oncology.

## 7.5 Enhancing theranostics via nanomedicine

To enhance the effectiveness of cancer treatment, integrating phytochemicals with nanotechnology offers a promising approach for both targeted therapy and diagnostics [144]. A novel strategy involves nanoformulating phytochemicals with fluorescent dyes, followed by conjugation to specific antibodies or peptides. This advanced combinational nanoformulation not only facilitates precise targeting of cancer cells but also enables real-time monitoring and diagnosis through fluorescence imaging [142].

The integration of phytochemicals into nanoformulations with fluorescent dyes and targeting agents, such as antibodies or peptides, allows for simultaneous therapeutic and diagnostic applications, commonly referred to as theranostics. This approach enables:

- *Targeted therapy*: by conjugating phytochemicals to antibodies or peptides that specifically bind to cancer cell markers, the nanoformulation can deliver therapeutic agents directly to tumor sites, minimizing off-target effects and enhancing treatment efficacy.

- *Real-time diagnostics*: the inclusion of fluorescent dyes provides a means for visualizing and tracking the distribution and accumulation of the nanoformulation in real time. This dual functionality helps in assessing treatment responses and adjusting therapeutic strategies accordingly.

Despite the significant potential benefits of these combinational nanoformulations, successful clinical translation requires addressing several challenges.

- *Cost and manufacturing*: the development of advanced nanoformulations can be expensive due to the complex processes involved in synthesizing and characterizing nanocarriers, fluorescent dyes, and targeting agents. Scaling up production while maintaining cost-effectiveness is crucial.

- *Safety and chronic toxicity*: ensuring the safety of these nanoformulations is paramount. Long-term studies are needed to assess the potential for chronic toxicity, especially with repeated exposure.

- *Batch-to-batch variability*: consistency in the quality and efficacy of nanoformulations across different production batches is essential for reliable clinical outcomes. Rigorous quality control measures and standardized manufacturing processes are required.

- *Regulatory approval*: navigating the regulatory landscape for novel nanoformulations involves comprehensive preclinical and clinical testing to demonstrate safety, efficacy, and compliance with regulatory standards.

Future research should focus on the following areas to overcome these obstacles and advance the clinical application of phytochemical-based nanoformulations:

- *Nanocarrier design optimization*: developing more efficient and cost-effective nanocarriers with enhanced targeting capabilities and reduced toxicity profiles.

- *Standardization of manufacturing processes*: implementing robust and reproducible manufacturing practices to ensure batch consistency and regulatory compliance.

- *Long-term safety studies*: conducting extensive preclinical and clinical trials to evaluate the long-term safety and efficacy of these nanoformulations.

- *Interdisciplinary collaboration*: encourage collaboration between researchers, clinicians, and industry professionals to address the multifaceted challenges associated with the development and clinical translation of these advanced therapeutic and diagnostic tools.

By addressing these challenges and focusing on strategic advancements, the integration of phytochemicals with nanotechnology holds the potential to revolutionize cancer treatment, offering more effective and personalized therapeutic options for patients [145].

## 8. Challenges and opportunities in phytochemical cancer therapy

Recent advances in phytochemical research have highlighted their potential in cancer therapy, with preclinical screening models assessing their toxicity, safety, pharmacokinetics, and efficacy. Several phytochemicals, including ursolic acid,

| Problem | Possible solution(s) |
|---|---|
| 1: Poor bioavailability of phytochemicals due to limited solubility, stability, and absorption rates, which restricts their effectiveness in reaching target tissues and exerting anticancer effects. | Improve bioavailability through nanoencapsulation, formulation optimization, and novel delivery systems to enhance pharmacokinetic profiles and therapeutic efficacy. |
| 2: Heterogeneity in phytochemical compositions caused by variations in plant sources, extraction methods, geographical origins, and cultivation conditions, leading to inconsistencies in phytochemical profiles, potency, and purity, affecting research reproducibility and clinical applications. | Standardize phytochemical extracts and formulations using rigorous quality control measures, including validated analytical techniques like HPLC, MS, and NMR spectroscopy. Establish standardized protocols for cultivation, harvesting, and extraction processes to minimize variability. |
| 3: Challenges in identifying and isolating active compounds within complex phytochemical mixtures due to synergistic or additive effects, which complicates understanding their therapeutic potential. | Conduct comprehensive studies using advanced techniques such as bioactivity-guided fractionation and systems biology approaches to identify key active constituents and understand their interactions. Perform robust pharmacokinetic and pharmacodynamic studies to optimize dosing regimens. |
| 4: Lack of mechanistic understanding of phytochemical actions in cancer cells, which hinders the rational design and optimization of treatments despite promising preclinical results. | Conduct further research to elucidate molecular mechanisms, identify biomarkers of response, and optimize treatment regimens based on patient characteristics and tumor biology. |
| 5: Challenges in translating preclinical findings into clinical practice due to regulatory approval, safety assessment, and clinical trial design issues, requiring rigorous methodologies and standardized protocols. | Systematically evaluate long-term safety data, potential drug interactions, and adverse effects to establish risk-benefit profiles and inform clinical decision-making. |
| 6: Limited awareness among healthcare professionals, variable insurance coverage, and cultural beliefs about natural remedies, which may affect patient access and acceptance of phytochemical-based treatments. | Integrate phytochemical-based therapies into conventional cancer care pathways through interdisciplinary collaboration, patient education, and healthcare provider training. |

*This table provides an overview of the main challenges encountered in developing phytochemical-based cancer treatments and corresponding approaches to address these issues.*

**Table 1.**
*Key challenges and proposed solutions in the development of phytochemical-based cancer treatments.*

baicalein, genistein, 6-shogaol, apigenin, thymoquinone, allicin, dicumarol, epigallocatechin, alpinumisoflavone, sulforaphane, curcumin, emodin, withaferin A, resveratrol, gingerol, physapubescin B, nimbolide, licochalcone A, glycyrrhizin, and hispidulin, have demonstrated preclinical efficacy using animal models [42]. However, most trials are in the early stages, with the most important phytochemicals under investigation targeting the autophagy-apoptosis pathway [142].

To optimize their clinical effectiveness, phytochemical-based cancer therapies must address several critical challenges (**Table 1**). One major issue is the poor bioavailability of these compounds, which is often due to their low solubility and absorption rates. Advanced strategies like nanoencapsulation and the development of novel delivery systems can mitigate this issue and enhance pharmacokinetics [137, 145]. Another challenge is the inherent variability in phytochemical compositions from different plant sources, which complicates efforts to standardize formulations for consistent dosage, efficacy, and safety [140]. Rigorous standardization, including the use of validated analytical methods and quality control measures, is essential to ensuring uniformity across batches. Identifying and isolating active compounds within complex phytochemical mixtures and understanding their synergistic effects requires sophisticated research techniques and comprehensive studies. Additionally, the limited mechanistic insight into phytochemical actions hampers the rational design and optimization of these treatments. Researchers are leveraging ethnobotanical knowledge, high-throughput screening, and Omics technologies to identify bioactive compounds and elucidate their mechanisms of action. Successful clinical translation necessitates thorough regulatory assessments, safety evaluations, and the creation of robust clinical trials. To fully integrate phytochemicals into conventional cancer care, it is crucial to foster interdisciplinary collaboration, enhance patient education, and address insurance and cultural barriers. Overcoming these challenges will enable phytochemical-based therapies to achieve their full potential as effective and safe cancer treatment options.

## 9. Final thoughts and future research directions

Cancer remains a significant challenge due to its rising prevalence and mortality rates, with conventional treatments like surgery and chemotherapy often facing limitations such as adverse effects, drug resistance, and lack of targeting precision. The complexity of cancer biology further complicates the development of effective, targeted therapies. Medicinal plants offer promising avenues for new pharmacological leads, supported by ethnopharmacological insights. Integrating this traditional knowledge with modern analytical and computational techniques can enhance drug discovery efforts. Phytochemicals, while showing potential for anti-tumorigenic effects by modulating key cellular and molecular pathways, face challenges related to stability, solubility, and metabolism. Nanomedical applications utilizing phytochemicals are emerging as a solution, improving drug delivery, enhancing therapeutic efficacy, and reducing side effects. These advances offer targeted strategies, particularly in combating cancer stem cells, and could significantly improve outcomes in cancer treatment.

The burgeoning epidemiological research on dietary patterns and cancer prevention highlights the need for tailored dietary recommendations, as no single "healthy" diet has been universally proven effective against all cancer types. While animal models offer insight into dietary impacts, replicating human diets and their effects remains challenging. Notably, dietary patterns like the ketogenic diet have been extensively studied in animals but require larger human studies to guide dietary guidelines. Furthermore, the question of personalizing dietary advice based on individual susceptibility factors such as genetics or microbiota remains unresolved. Phytochemicals, known for their anticancer properties through mechanisms like apoptosis induction and oxidative stress inhibition, have shown promise, but most studies are limited by *in vitro* models and ethnopharmacological data. The variability in phytochemical effects and potential negative impacts complicates their application. Despite the encouraging preclinical results, issues with bioavailability and toxicity underscore the need for cautious, well-designed human clinical trials to better understand the efficacy, safety, and interactions of phytochemicals as cancer prevention agents.

Phytochemicals, though recognized for their potential health benefits and protective effects, present challenges in determining optimal intake levels due to their diverse biological activities and lack of established disease associations. Efforts to establish consumption guidelines are complicated by factors such as bioavailability, food processing, and individual variability in absorption. For instance, a US study demonstrated that 13–22 g of dietary fiber daily can reduce cholesterol by 7%, while recommended intake levels for carotenoids vary significantly across populations. Despite the anticancer potential of phytochemicals and supportive epidemiological data, clinical trials often fail to replicate these benefits, sometimes due to early termination or safety concerns. Specific phytochemicals, such as capsaicin, have shown both protective and potentially carcinogenic effects, underscoring the need for careful consumption and further research. Preclinical studies suggest promising therapeutic potential, but more in-depth molecular and clinical investigations are needed to fully understand their effects and ensure safety. *In silico* tools like molecular docking, combined with *in vitro* and *in vivo* studies, are crucial for elucidating these interactions and advancing phytochemical research.

Phytochemical-based nanotechnology holds significant promise for overcoming the limitations of conventional anticancer drugs, such as low solubility, poor specificity, and short half-life. By combining phytochemicals with synthetic drugs in nanoparticle-based delivery systems, bioavailability can be enhanced, toxicity reduced, and circulation time extended. Nanotechnology can improve the solubility, stability, and targeted delivery of these compounds, although challenges remain, such as limited encapsulation capacity and manufacturing complexities. Further research is needed to optimize these systems for clinical use. Nanoparticle-based drug delivery systems (NDDSs) are emerging as key tools for cancer diagnostics, imaging, and treatment, with potential applications in early diagnosis, targeted therapy, and mitigating mitochondrial impairments. Tailoring these approaches to individual patient profiles and integrating them with conventional treatments could significantly advance cancer care.

The pharmaceutical industry is increasingly integrating traditional and modern approaches to enhance drug development, particularly through the combination of plant-derived phytochemicals with conventional chemotherapeutics. This strategy aims to leverage the synergistic effects of these combinations, which have shown promise in improving therapeutic efficiency and minimizing side effects in cancer treatment. Recent research highlights the potential of phytochemicals to target cancer cells more specifically while sparing normal cells, thanks to their low intrinsic toxicity. Nanotechnology-based delivery systems, including solid lipid nanoparticles, liposomes, and polymeric micelles, are enhancing the efficacy of these phytochemical-drug combinations by improving solubility, targeting, and bioavailability. However, optimizing these combinations remains a challenge due to the need for precise drug exposure sequences and ratios. Advanced pharmacodynamic evaluation is crucial for justifying the use of these nanoformulations. Preliminary studies suggest that combinations of phytochemicals, such as lycopene and compounds from carrots, can affect multiple cancer-related pathways and improve treatment outcomes. Nonetheless, further research is necessary to fully understand their interactions, potential side effects, and clinical applicability.

Autophagy and apoptosis are critical cellular processes in cancer development and treatment, with their interplay influencing cancer progression and drug resistance. Defects in apoptotic signaling often lead to chemotherapy resistance, prompting the exploration of autophagy as an alternative mechanism for cancer therapy. Further investigations are needed to better understand cancer pathobiology and exploit the full potential of autophagy-apoptosis-targeted drug design. Phytochemicals such as sulforaphane, resveratrol, lycopene, epigallocatechin, curcumin, and berberine, known for their selective anticancer properties, modulate both autophagy and apoptosis. While these compounds show promise, their multi-target effects and the complexity of cellular signaling pathways present challenges in drug development. To optimize phytochemical-based therapies, further research is needed to understand their mechanisms through *in vitro* and *in vivo* studies, and by integrating system pharmacology and computational approaches to better elucidate their anticancer potential.

Clinical studies on phytochemicals often face methodological challenges, such as inadequate control groups, small sample sizes, and brief trial durations. To establish the efficacy, safety, and adverse effects of phytochemicals in cancer treatment, large-scale, well-controlled clinical trials are essential. There is a critical need for standardized methods across bioavailability, efficacy, safety, quality, and regulatory practices to meet international standards. While extensive research on cell lines and animal models exists, human studies on phytochemical benefits remain limited. Additionally, evidence suggests that phytochemicals from fruits and vegetables may offer benefits for gastrointestinal cancers and improve prognosis. Further research is required to validate these findings and explore the role of bacteria in metabolizing phytochemicals. Overall, a diet rich in fruits and vegetables is recommended for its potential to enhance cancer prognosis and possibly aid in prevention.

Future research on phytochemicals for cancer therapy will harness advances in biotechnology and molecular biology to clarify their mechanisms and establish clinical efficacy. Key efforts will involve using high-throughput screening, genomics, proteomics, and metabolomics to understand how phytochemicals interact with cellular targets and signaling pathways. Improving their bioavailability through advanced delivery systems like nanoparticles and liposomes is essential for enhancing clinical effectiveness. Rigorous clinical trials, including large-scale randomized controlled trials with extended follow-ups, will be crucial for validating safety, efficacy, and optimal dosing. Additionally, exploring synergistic effects with conventional treatments and integrating phytochemicals into cancer prevention strategies could open new therapeutic and preventive possibilities. Ensuring public and regulatory acceptance, supported by education and robust regulatory frameworks, will be critical for the successful adoption of phytochemical-based therapies.

## Acknowledgements

The author would like to express sincere thanks to Chief MOSOH Paul Tandong and Chieftess Ateyim Espe MOSOH Ostensia Nkeng of PINYIN (Santa, North-West Region, Cameroon) for their unwavering and invaluable support. Many thanks to Mr. Tetu Acha Samuel (MD, USA) and Mr. Adamou Musa (TX, USA) for their support.

In preparing this work, the author utilized Quillbot to polish the manuscript's language. After using this service, the author carefully reviewed and edited the content as necessary, assuming full responsibility for the publication's content.

## Conflict of interest

The author states that no known competing financial interest or personal relationship influenced the work described in this publication.

## Author details

Dexter Achu Mosoh[1,2,3]

1 Centre for Biodiversity Exploration and Conservation (CBEC), Jabalpur, MP, India

2 School of Sciences, Sanjeev Agrawal Global Educational (SAGE) University, Bhopal, MP, India

3 Prof. Wagner A. Vendrame's Laboratory, Environmental Horticulture Department, University of Florida, Institute of Food and Agricultural Sciences, Gainesville, FL, USA

*Address all correspondence to: mosohdexter@hotmail.com

IntechOpen

© 2024 The Author(s). Licensee IntechOpen. This chapter is distributed under the terms of the Creative Commons Attribution License (http://creativecommons.org/licenses/by/4.0), which permits unrestricted use, distribution, and reproduction in any medium, provided the original work is properly cited.

## References

[1] Bray F et al. Global cancer statistics 2022: GLOBOCAN estimates of incidence and mortality worldwide for 36 cancers in 185 countries. CA: A Cancer Journal for Clinicians. 2024;**74**(3):229-263

[2] Mattiuzzi C, Lippi G. Current cancer epidemiology. Journal of Epidemiology and Global Health. 2019;**9**(4):217. DOI: 10.2991/jegh.k.191008.001

[3] Montagnana M, Lippi G. Cancer diagnostics: Current concepts and future perspectives. Annals of Translational Medicine. 2017;**5**(13):268-268. DOI: 10.21037/atm.2017.06.20

[4] Statista. Projection on Cancer Care Costs U.S. 2030. Statista. Available from: https://www.statista.com/statistics/1313989/projection-on-cancer-care-costs-us/ [Accessed: June 01, 2024]

[5] NCI. Common Cancer Sites - Cancer Stat Facts. SEER. Available from: https://seer.cancer.gov/statfacts/html/common.html [Accessed: June 01, 2024]

[6] Pruller J. Intratumoural heterogeneity as a major challenge for cancer modelling and successful treatment. Science Reviews - Biology. 2023;**2**(1):12-19. DOI: 10.57098/SciRevs.Biology.2.1.2

[7] Bhatia S, Frangioni JV, Hoffman RM, Iafrate AJ, Polyak K. The challenges posed by cancer heterogeneity. Nature Biotechnology. 2012;**30**(7):604-610. DOI: 10.1038/nbt.2294

[8] Ottaiano A et al. From chaos to opportunity: Decoding cancer heterogeneity for enhanced treatment strategies. Biology. 2023;**12**(9):1183. DOI: 10.3390/biology12091183

[9] Chew AH, Chandramouli N, Kanesvaran R, Iyer GN, Hwang WY. Winning the fight against cancer. Annals of the Academy of Medicine, Singapore. 2020;**49**(10):779-788. DOI: 10.47102/annals-acadmedsg.2020204

[10] Malafi ME. The lives saved: A literature review on the role of radiotherapy improving prognosis in cancer patients. Journal of Quality in Health Care & Economics. 2024;**7**(1):1-3. DOI: 10.23880/jqhe-16000360

[11] Zhou Z, Li M. Targeted therapies for cancer. BMC Medicine. 2022;**20**(1):90. DOI: 10.1186/s12916-022-02287-3

[12] Hoda M. Potential alternatives to conventional cancer therapeutic approaches: The way forward. Current Pharmaceutical Biotechnology. 2021;**22**(9):1141-1148. DOI: 10.2174/1389201021666201016142408

[13] Maeda H, Khatami M. Analyses of repeated failures in cancer therapy for solid tumors: Poor tumor-selective drug delivery, low therapeutic efficacy and unsustainable costs. Clinical and Translational Medicine. 2018;**7**(1):e11. DOI: 10.1186/s40169-018-0185-6

[14] Latha N, Komahan G. Mathematical model in side effects of chemotherapy treatment for lung cancer. International Journal of Innovative Science and Research Technology. 2017;**3**(10):267-270

[15] Madhavi A, Reddy D, Basha DS, Hussain S, Afreen S, Haritha S. A review on chemotherapy induced complications in cancer patients. World Journal of Current Medical and Pharmaceutical Research. 2019. Available from: https://

www.semanticscholar.org/paper/A-REVIEW-ON-CHEMOTHERAPY-INDUCED-COMPLICATIONS-IN-Madhavi-Reddy/cd81113254937e4ac2792cd6b2699d191d0b95ea [Accessed: June 03, 2024]

[16] Van Den Boogaard WMC, Komninos DSJ, Vermeij WP. Chemotherapy side-effects: Not all DNA damage is equal. Cancers. 2022;**14**(3):627. DOI: 10.3390/cancers14030627

[17] Basha S, Sk H, Sk SA, Haritha S. A review on chemotherapy induced complications in cancer patients. World Journal of Current Medical and Pharmaceutical Research. 2019;**1**(6):216-222. Available from: https://wjcmpr.com/index.php/journal/article/view/63

[18] Adashek JJ, Goloubev A, Kato S, Kurzrock R. Missing the target in cancer therapy. Nature Cancer. 2021;**2**(4):369-371. DOI: 10.1038/s43018-021-00204-w

[19] Barkat HA, Das SS, Barkat MA, Beg S, Hadi HA. Selective targeting of cancer signaling pathways with nanomedicines: Challenges and progress. Future Oncology. 2020;**16**(35):2959-2979. DOI: 10.2217/fon-2020-0198

[20] Bastien J-P, Minguy A, Dave V, Roy DC. Cellular therapy approaches harnessing the power of the immune system for personalized cancer treatment. Seminars in Immunology. 2019;**42**:101306. DOI: 10.1016/j.smim.2019.101306

[21] Balzerano A, Paccosi E, Proietti-De-Santis L. Evolutionary mechanisms of cancer suggest rational therapeutic approaches. Cytogenetic and Genome Research. 2021;**161**(6-7):362-371. DOI: 10.1159/000516530

[22] Krzyszczyk P et al. The growing role of precision and personalized medicine for cancer treatment. Technology. 2018;**6**(3-4):79-100. DOI: 10.1142/S2339547818300020

[23] Schade S et al. A data- and model-driven approach for cancer treatment. Onkologie. 2019;**25**(S2):132-137. DOI: 10.1007/s00761-019-0624-z

[24] Vaseghi Maghvan P et al. Personalized medicine in colorectal cancer. Gastroenterology and Hepatology From Bed to Bench. 2020;**13**(Suppl. 1):S18-S28

[25] Goetz LH, Schork NJ. Personalized medicine: Motivation, challenges and progress. Fertility and Sterility. 2018;**109**(6):952-963. DOI: 10.1016/j.fertnstert.2018.05.006

[26] Shegai PV, Shatalov PA, Zabolotneva AA, Falaleeva NA, Ivanov SA, Kaprin AD. Challenges faced by clinicians in the personalized treatment planning: A literature review and the first results of the Russian National Cancer Program. Critical Care Research and Practice. 2021;**2021**:6649771. DOI: 10.1155/2021/6649771

[27] Saeed RF, Awan UA, Saeed S, Mumtaz S, Akhtar N, Aslam S. Targeted therapy and personalized medicine. Cancer Treatment and Research. 2023;**185**:177-205. DOI: 10.1007/978-3-031-27156-4_10

[28] Billaud A, Chevalier L-M, Campone M, Morel A, Bigot F. L'instabilité génomique, paramètre limitant l'efficacité des thérapies ciblées en oncologie. Bulletin du Cancer (Paris). 2020;**107**(11):1161-1170. DOI: 10.1016/j.bulcan.2020.08.010

[29] Corso S, Giordano S. Mechanisms of resistance to molecular therapies targeting the HGF/MET axis. In:

Yarden Y, Elkabets M, editors. Resistance to Anti-Cancer Therapeutics Targeting Receptor Tyrosine Kinases and Downstream Pathways. Cham: Springer International Publishing; 2018. pp. 67-87. DOI: 10.1007/978-3-319-67932-7_4

[30] Wajapeyee N, Gupta R. Epigenetic alterations and mechanisms that drive resistance to targeted cancer therapies. Cancer Research. 2021;**81**(22):5589-5595. DOI: 10.1158/0008-5472.CAN-21-1606

[31] Montoya S, Soong D, Nguyen N, Affer M, Munamarty SP, Taylor J. Targeted therapies in cancer: To be or not to be, selective. Biomedicine. 2021;**9**(11):Art. no. 11. DOI: 10.3390/biomedicines9111591

[32] Mosoh DA. Widely-targeted in silico and in vitro evaluation of veratrum alkaloid analogs as FAK inhibitors and dual targeting of FAK and Hh/SMO pathways for cancer therapy: A critical analysis. International Journal of Biological Macromolecules. 2024;**281**:136201. DOI: 10.1016/j.ijbiomac.2024.136201

[33] Kumar A, Singh K, Kumar K, Kumar S, Singh A. Immunotherapy in cancer treatment: Harnessing the power of the immune system. Research Journal of Pharmaceutical Dosage Forms and Technology. 2024;**16**(1):107-112. DOI: 10.52711/0975-4377.2024.00017

[34] Gerson JN, Ramamurthy C, Borghaei H. Managing adverse effects of immunotherapy. Clinical Advances in Hematology & Oncology : H&O. 2018;**16**(5):364-374

[35] Swami U, Milhem M. Cancer immunotherapy: Recent advances and challenges. Annals of Translational Medicine. 2021;**9**(12):1032. DOI: 10.21037/atm-21-2325

[36] Ventola CL. Cancer immunotherapy, part 3: Challenges and future trends. Pharmacology & Therapeutics. 2017;**42**(8):514-521

[37] Geethakumari PR, Ramasamy DP, Dholaria B, Berdeja J, Kansagra A. Balancing quality, cost, and access during delivery of newer cellular and immunotherapy treatments. Current Hematologic Malignancy Reports. 2021;**16**(4):345-356. DOI: 10.1007/s11899-021-00635-3

[38] Ramamurthy D et al. Recent advances in immunotherapies against infectious diseases. Immunotherapy Advances. 2021;**1**(1):ltaa007

[39] Koh Y-C, Ho C-T, Pan M-H. Recent advances in cancer chemoprevention with phytochemicals. Journal of Food and Drug Analysis. 2020;**28**(1):14-37. DOI: 10.1016/j.jfda.2019.11.001

[40] Cianciosi D et al. Targeting molecular pathways in cancer stem cells by natural bioactive compounds. Pharmacological Research. 2018;**135**:150-165

[41] Kaur V, Kumar M, Kumar A, Kaur K, Dhillon VS, Kaur S. Pharmacotherapeutic potential of phytochemicals: Implications in cancer chemoprevention and future perspectives. Biomedicine & Pharmacotherapy. 2018;**97**:564-586. DOI: 10.1016/j.biopha.2017.10.124

[42] Choudhari AS, Mandave PC, Deshpande M, Ranjekar P, Prakash O. Phytochemicals in cancer treatment: From preclinical studies to clinical practice. Frontiers in Pharmacology. 2020;**10**:1614. DOI: 10.3389/fphar.2019.01614

[43] Ibrahim SM, Ali M, Aminu F, Mu'azu L. Potential phytochemicals for cancer

treatment: A review. Journal of Oncology Research. 2021;**4**(1):9. DOI: 10.30564/jor.v4i1.3656

[44] Mouhid L et al. Improving in vivo efficacy of bioactive molecules: An overview of potentially antitumor phytochemicals and currently available lipid-based delivery systems. Journal of Oncology. 2017;**2017**(1):7351976. DOI: 10.1155/2017/7351976

[45] Garcia-Oliveira P et al. Status and challenges of plant-anticancer compounds in cancer treatment. Pharmaceuticals. 2021;**14**(2):157. DOI: 10.3390/ph14020157

[46] Mosoh DA, Khandel AK, Verma SK, Vendrame WA. Effects of sterilization methods and plant growth regulators on in vitro regeneration and tuberization in *Gloriosa superba* (L.). In Vitro Cellular & Developmental Biology - Plant. 2023;**59**:792807. DOI: 10.1007/s11627-023-10387-9

[47] Mosoh DA, Khandel AK, Verma SK, Vendrame WA. Phytochemical analysis and enhanced production of alkaloids in non-dormant corm-derived callus of *Gloriosa superba* (L.) using plant growth regulators and abiotic elicitors. Plant Cell, Tissue, and Organ Cultures. PCTOC. 2024;**156**(3):89. DOI: 10.1007/s11240-023-02674-5

[48] Mosoh DA, Khandel AK, Verma SK, Vendrame WA. Standardizing in vitro callus induction and indirect organogenesis of Gloriosa superba L. leaf explants using exogenous phytohormones. Journal of Plant Biotechnology. 2024;**51**:237-252. DOI: 10.5010/JPB.2024.51.023.237

[49] Kapinova A et al. Are plant-based functional foods better choice against cancer than single phytochemicals? A critical review of current breast cancer research. Biomedicine & Pharmacotherapy. 2017;**96**:1465-1477

[50] Shanmugam MK et al. Cancer prevention and therapy through the modulation of transcription factors by bioactive natural compounds. Seminars in Cancer Biology. 2016;**40**:35-47. DOI: 10.1016/j.semcancer.2016.03.005

[51] Ullah S, et al. Phytochemicals for cancer treatment: An update on plant-derived anti-cancer compounds and their mechanisms of action. Current Topics in Medicinal Chemistry. 2024. DOI: 10.2174/0115680266268185240320085619

[52] Iqbal J et al. Potential phytochemicals in the fight against skin cancer: Current landscape and future perspectives. Biomedicine & Pharmacotherapy. 2019;**109**:1381-1393. DOI: 10.1016/j.biopha.2018.10.107

[53] Mani J et al. Bioassay guided fractionation protocol for determining novel active compounds in selected Australian Flora. Plants. 2022;**11**(21):Art. no. 21. DOI: 10.3390/plants11212886

[54] Phillipson DW et al. High-throughput bioassay-guided fractionation: A technique for rapidly assigning observed activity to individual components of combinatorial libraries, screened in HTS bioassays. Journal of Combinatorial Chemistry. 2002;**4**(6):591-599. DOI: 10.1021/cc020042e

[55] Wang H et al. Plants against cancer: A review on natural phytochemicals in preventing and treating cancers and their druggability. Anti-Cancer Agents in Medicinal Chemistry. 2012;**12**(10):1281-1305

[56] Hosseini A, Ghorbani A. Cancer therapy with phytochemicals: Evidence from clinical studies. Avicenna Journal of Phytomedicine.

2015. Available from: https://www.semanticscholar.org/paper/Cancer-therapy-with-phytochemicals%3A-evidence-from-Hosseini-Ghorbani/9650572ff0a440d895fabf4cf1cc88b57329d993 [Accessed: June 06, 2024]

[57] Singh CK, Ndiaye MA, Ahmad N. Resveratrol and cancer: Challenges for clinical translation. Biochimica et Biophysica Acta (BBA) - Molecular Basis of Disease. 1852;**2015**(6):1178-1185. DOI: 10.1016/j.bbadis.2014.11.004

[58] Singh VK, Arora D, Ansari MI, Sharma PK. Phytochemicals based chemopreventive and chemotherapeutic strategies and modern technologies to overcome limitations for better clinical applications. Phytotherapy Research. 2019;**33**(12):3064-3089. DOI: 10.1002/ptr.6508

[59] Brown K, Rufini A. New concepts and challenges in the clinical translation of cancer preventive therapies: The role of pharmacodynamic biomarkers. Ecancermedicalscience. 2015;**9**:601. DOI: 10.3332/ecancer.2015.601

[60] Aqil F, Munagala R, Jeyabalan J, Vadhanam MV. Bioavailability of phytochemicals and its enhancement by drug delivery systems. Cancer Letters. 2013;**334**(1):133-141. DOI: 10.1016/j.canlet.2013.02.032

[61] Yim CY, Mao P, Spinella MJ. Headway and hurdles in the clinical development of dietary phytochemicals for cancer therapy and prevention: Lessons learned from vitamin A derivatives. The AAPS Journal. 2014;**16**(2):281-288. DOI: 10.1208/s12248-014-9562-2

[62] Bollag W, Holdener EE. Retinoids in cancer prevention and therapy. Annals of Oncology. 1992;**3**(7):513-526. DOI: 10.1093/oxfordjournals.annonc.a058252

[63] Beuth J, Schierholz JM. Review. Evidence-based complementary oncology. Innovative approaches to optimize standard therapy strategies. Vivo Athens Greece. 2007;**21**(2):423-428

[64] Dutt R, Garg V, Khatri N, Madan AK. Phytochemicals in anticancer drug development. Anti-Cancer Agents in Medicinal Chemistry (Formerly Current Medicinal Chemistry - Anti-cancer Agents). 2019;**19**(2):172-183

[65] Mosoh DA, Khandel AK, Verma SK, Vendrame WA. Optimizing callus induction and indirect organogenesis in non-dormant corm explants of *Gloriosa superba* (L.) via media priming. Frontiers in Horticulture. 2024;**3**:1378098. DOI: 10.3389/fhort.2024.1378098

[66] Mosoh DA, Khandel AK, Verma SK, Vendrame WA. Overcoming dual seed dormancy and enhancing in vitro seedling development of *Gloriosa superba* (L.) with a targeted sterilization approach and plant growth regulator synergy. Tropical Plants. 2024;**3**:e031. DOI: 10.48130/tp-0024-0033

[67] Mosoh DA, Prakash O, Khandel AK, Vendrame WA. Preserving Earth's flora in the 21st century: Climate, biodiversity, and Global Change Factors (GCFs) since the mid-1940s. Frontiers in Conservation Science. 2024;**5**:1383370. DOI: 10.3389/fcosc.2024.1383370

[68] Mosoh DA, Khandel AK, Verma SK, Vendrame WA. Multi-explant and multiplex applications of plant growth regulators: A critical analysis of direct organogenesis in Gloriosa superba (L.). Tropical Plants. 2024. DOI: 10.48130/tp-0024-0038

[69] Rudzińska A et al. Phytochemicals in cancer treatment and cancer prevention—Review on epidemiological

data and clinical trials. Nutrients. 2023;**15**(8):Art. no. 8. DOI: 10.3390/nu15081896

[70] Ruiz RB, Hernández PS. Cancer chemoprevention by dietary phytochemicals: Epidemiological evidence. Maturitas. 2016;**94**:13-19

[71] Steck SE, Murphy EA. Dietary patterns and cancer risk. Nature Reviews. Cancer. 2020;**20**(2):125-138. DOI: 10.1038/s41568-019-0227-4

[72] Estrela JM et al. Polyphenolic phytochemicals in cancer prevention and therapy: Bioavailability versus bioefficacy. Journal of Medicinal Chemistry. 2017;**60**(23):9413-9436. DOI: 10.1021/acs.jmedchem.6b01026

[73] Russo GL, Spagnuolo C, Russo M. Reassessing the role of phytochemicals in cancer chemoprevention. Biochemical Pharmacology. 2024;**228**:116165. DOI: 10.1016/j.bcp.2024.116165

[74] Woźniak M, Krajewski R, Makuch S, Agrawal S. Phytochemicals in gynecological cancer prevention. International Journal of Molecular Sciences. 2021;**22**(3):Art. no. 3. DOI: 10.3390/ijms22031219

[75] Shenoy G. When little things make a big difference. Immunological Investigations. 2020;**49**(7):692-697. DOI: 10.1080/08820139.2020.1831774

[76] Lee AH, Dixit VD. Dietary regulation of immunity. Immunity. 2020;**53**(3):510-523. DOI: 10.1016/j.immuni.2020.08.013

[77] Pinho SS, Alves I, Gaifem J, Rabinovich GA. Immune regulatory networks coordinated by glycans and glycan-binding proteins in autoimmunity and infection. Cellular & Molecular Immunology. 2023;**20**(10):1101-1113. DOI: 10.1038/s41423-023-01074-1

[78] Rabinovich GA, van Kooyk Y, Cobb BA. Glycobiology of immune responses. Annals of the New York Academy of Sciences. 2012;**1253**(1):1-15. DOI: 10.1111/j.1749-6632.2012.06492.x

[79] Landis-Piwowar KR. Overview of the immune response and regulation. American Society for Clinical Laboratory Science. 2015;**28**(1):35-37. DOI: 10.29074/ascls.28.1.35

[80] Hinshaw DC, Shevde LA. The tumor microenvironment innately modulates cancer progression. Cancer Research. 2019;**79**(18):4557-4566. DOI: 10.1158/0008-5472.CAN-18-3962

[81] Zamarron BF, Chen W. Dual roles of immune cells and their factors in cancer development and progression. International Journal of Biological Sciences. 2011;**7**(5):651-658. DOI: 10.7150/ijbs.7.651

[82] Trac NT, Chung EJ. Peptide-based targeting of immunosuppressive cells in cancer. Bioactive Materials. 2020;**5**(1):92-101. DOI: 10.1016/j.bioactmat.2020.01.006

[83] Malla RR, Deepak K, Merchant N, Dasari VR. Breast tumor microenvironment: Emerging target of therapeutic phytochemicals. Phytomedicine. 2020;**70**:153227. DOI: 10.1016/j.phymed.2020.153227

[84] Renner K et al. Metabolic hallmarks of tumor and immune cells in the tumor microenvironment. Frontiers in Immunology. 2017;**8**:248. DOI: 10.3389/fimmu.2017.00248

[85] Albahri G et al. Potential anti-tumorigenic properties of diverse medicinal plants against the majority of common types of cancer. Pharmaceuticals. 2024;**17**(5):Art. no. 5. DOI: 10.3390/ph17050574

[86] Paudel S, Mishra N, Agarwal R. Phytochemicals as immunomodulatory molecules in cancer therapeutics. Pharmaceuticals. 2023;**16**(12):Art. no. 12. DOI: 10.3390/ph16121652

[87] Rana P, Shrama A, Mandal CC. Molecular insights into phytochemicals-driven break function in tumor microenvironment. Journal of Food Biochemistry. 2021;**45**(9):e13824. DOI: 10.1111/jfbc.13824

[88] Zhang Q, Yang C, Gao X, Dong J, Zhong C. Phytochemicals in regulating PD-1/PD-L1 and immune checkpoint blockade therapy. Phytotherapy Research. 2024;**38**(2):776-796. DOI: 10.1002/ptr.8082

[89] Lei Y et al. Phytochemical natural killer cells reprogram tumor microenvironment for potent immunotherapy of solid tumors. Biomaterials. 2022;**287**:121635. DOI: 10.1016/j.biomaterials.2022.121635

[90] Yin S-Y, Yang N-S, Lin T-J. Phytochemicals approach for developing cancer immunotherapeutics. Frontiers in Pharmacology. 2017;**8**:386. DOI: 10.3389/fphar.2017.00386

[91] Singh RP, Agarwal R. Tumor angiogenesis: A potential target in cancer control by phytochemicals. Current Cancer Drug Targets. 2003;**3**(3):205-217. DOI: 10.2174/1568009033481985

[92] Lee J et al. Phytochemicals in cancer immune checkpoint inhibitor therapy. Biomolecules. 2021;**11**(8):Art. no. 8. DOI: 10.3390/biom11081107

[93] Catalano E. Role of phytochemicals in the chemoprevention of tumors. arXiv preprint arXiv:1605.04519. 2016. DOI: 10.48550/arXiv.1605.04519

[94] Castañeda AM, Meléndez CM, Uribe D, Pedroza-Díaz J. Synergistic effects of natural compounds and conventional chemotherapeutic agents: Recent insights for the development of cancer treatment strategies. Heliyon. 2022;**8**(6):e09519. DOI: 10.1016/j.heliyon.2022.e09519

[95] Farzaei MH, Bahramsoltani R, Rahimi R. Phytochemicals as adjunctive with conventional anticancer therapies. Current Pharmaceutical Design. 2016;**22**(27):4201-4218. DOI: 10.2174/1381612822666160601100823

[96] Liu Y-Q, Wang X-L, He D-H, Cheng Y-X. Protection against chemotherapy- and radiotherapy-induced side effects: A review based on the mechanisms and therapeutic opportunities of phytochemicals. Phytomedicine. 2021;**80**:153402. DOI: 10.1016/j.phymed.2020.153402

[97] Sak K. Chemotherapy and dietary phytochemical agents. Chemotherapy Research and Practice. 2012;**2012**(1):282570. DOI: 10.1155/2012/282570

[98] Jain A, Madu CO, Lu Y. Phytochemicals in chemoprevention: A cost-effective complementary approach. Journal of Cancer. 2021;**12**(12):3686-3700. DOI: 10.7150/jca.57776

[99] Ranjan A, Fofaria NM, Kim S-H, Srivastava SK. Modulation of signal transduction pathways by natural compounds in cancer. Chinese Journal of Natural Medicines. 2015;**13**(10):730-742

[100] Das B, Sarkar N, Bishayee A, Sinha D. Dietary phytochemicals in the regulation of epithelial to mesenchymal transition and associated enzymes: A promising anticancer therapeutic approach. Seminars in Cancer Biology. 2019;**56**:196-218. DOI: 10.1016/j.semcancer.2018.11.007

[101] Wang S, Shen P, Zhou J, Lu Y. Diet phytochemicals and cutaneous carcinoma chemoprevention: A review. Pharmacological Research. 2017;**119**:327-346

[102] Sayeed MA, Bracci M, Lucarini G, Lazzarini R, Di Primio R, Santarelli L. Regulation of microRNA using promising dietary phytochemicals: Possible preventive and treatment option of malignant mesothelioma. Biomedicine & Pharmacotherapy. 2017;**94**:1197-1224

[103] Reuter S, Gupta SC, Chaturvedi MM, Aggarwal BB. Oxidative stress, inflammation, and cancer: How are they linked? Free Radical Biology & Medicine. 2010;**49**(11):1603-1616. DOI: 10.1016/j.freeradbiomed.2010.09.006

[104] Liu T, Zhang L, Joo D, Sun S-C. NF-κB signaling in inflammation. Signal Transduction and Targeted Therapy. 2017;**2**(1):1-9. DOI: 10.1038/sigtrans.2017.23

[105] Sethi G, Sung B, Aggarwal BB. TNF: A master switch for inflammation to cancer. Frontiers in Bioscience-Landmark. 2008;**13**(13):Art. no. 13. DOI: 10.2741/3066

[106] Hoesel B, Schmid JA. The complexity of NF-κB signaling in inflammation and cancer. Molecular Cancer. 2013;**12**(1):86. DOI: 10.1186/1476-4598-12-86

[107] Jiang J, Dingledine R. Prostaglandin receptor EP2 in the crosshairs of anti-inflammation, anti-cancer, and neuroprotection. Trends in Pharmacological Sciences. 2013;**34**(7):413-423. DOI: 10.1016/j.tips.2013.05.003

[108] Yamamoto Y, Gaynor RB. Role of the NF-kB pathway in the pathogenesis of human disease states. Current Molecular Medicine. 2001;**1**(3):287-296. DOI: 10.2174/1566524013363816

[109] Chikara S, Nagaprashantha LD, Singhal J, Horne D, Awasthi S, Singhal SS. Oxidative stress and dietary phytochemicals: Role in cancer chemoprevention and treatment. Cancer Letters. 2018;**413**:122-134. DOI: 10.1016/j.canlet.2017.11.002

[110] Li A-N, Li S, Zhang Y-J, Xu X-R, Chen Y-M, Li H-B. Resources and biological activities of natural polyphenols. Nutrients. 2014;**6**(12):Art. no. 12. DOI: 10.3390/nu6126020

[111] Lee JH, Khor TO, Shu L, Su Z-Y, Fuentes F, Kong A-NT. Dietary phytochemicals and cancer prevention: Nrf2 signaling, epigenetics, and cell death mechanisms in blocking cancer initiation and progression. Pharmacology & Therapeutics. 2013;**137**(2):153-171. DOI: 10.1016/j.pharmthera.2012.09.008

[112] Chen F et al. Different inhibition of Nrf2 by two Keap1 isoforms α and β to shape malignant behaviour of human hepatocellular carcinoma. International Journal of Molecular Sciences. 2022;**23**(18):Art. no. 18. DOI: 10.3390/ijms231810342

[113] Jaramillo MC, Zhang DD. The emerging role of the Nrf2–Keap1 signaling pathway in cancer. Genes & Development. 2013;**27**(20):2179-2191. DOI: 10.1101/gad.225680.113

[114] Braicu C et al. Natural compounds modulate the crosstalk between apoptosis- and autophagy-regulated signaling pathways: Controlling the uncontrolled expansion of tumor cells. Seminars in Cancer Biology. 2022;**80**:218-236. DOI: 10.1016/j.semcancer.2020.05.015

[115] Wani AK et al. Targeting apoptotic pathway of cancer cells with phytochemicals and plant-based nanomaterials. Biomolecules. 2023;**13**(2):194. DOI: 10.3390/biom13020194

[116] Cháirez-Ramírez MH, de la Cruz-López KG, García-Carrancá A. Polyphenols as antitumor agents targeting key players in cancer-driving signaling pathways. Frontiers in Pharmacology. 2021;**12**:710304. DOI: 10.3389/fphar.2021.710304

[117] Delmas D, Passilly-Degrace P, Jannin B, Malki MC, Latruffe N. Resveratrol, a chemopreventive agent, disrupts the cell cycle control of human SW480 colorectal tumor cells. International Journal of Molecular Medicine. 2002;**10**(2):193-199. DOI: 10.3892/ijmm.10.2.193

[118] Hwang E-S, Bowen PE. Cell cycle arrest and induction of apoptosis by lycopene in LNCaP human prostate cancer cells. Journal of Medicinal Food. 2004;**7**(3):284-289. DOI: 10.1089/jmf.2004.7.284

[119] Hasima N, Ozpolat B. Regulation of autophagy by polyphenolic compounds as a potential therapeutic strategy for cancer. Cell Death & Disease. 2014;**5**(11):e1509-e1509. DOI: 10.1038/cddis.2014.467

[120] Stacchiotti A, Corsetti G. Natural compounds and autophagy: Allies against neurodegeneration. Frontiers in Cell and Development Biology. 2020;**8**:555409. DOI: 10.3389/fcell.2020.555409

[121] Li X, He S, Ma B. Autophagy and autophagy-related proteins in cancer. Molecular Cancer. 2020;**19**(1):12. DOI: 10.1186/s12943-020-1138-4

[122] Patra S et al. Autophagy-modulating phytochemicals in cancer therapeutics: Current evidences and future perspectives. Seminars in Cancer Biology. 2022;**80**:205-217. DOI: 10.1016/j.semcancer.2020.05.008

[123] Al-Ishaq RK, Overy AJ, Büsselberg D. Phytochemicals and gastrointestinal cancer: Cellular mechanisms and effects to change cancer progression. Biomolecules. 2020;**10**(1):Art. no. 1. DOI: 10.3390/biom10010105

[124] Schumacher M, Kelkel M, Dicato M, Diederich M. Gold from the sea: Marine compounds as inhibitors of the hallmarks of cancer. Biotechnology Advances. 2011;**29**(5):531-547. DOI: 10.1016/j.biotechadv.2011.02.002

[125] Afrin S et al. Dietary phytochemicals in colorectal cancer prevention and treatment: A focus on the molecular mechanisms involved. Biotechnology Advances. 2020;**38**:107322. DOI: 10.1016/j.biotechadv.2018.11.011

[126] Schnekenburger M, Dicato M, Diederich M. Plant-derived epigenetic modulators for cancer treatment and prevention. Biotechnology Advances. 2014;**32**(6):1123-1132. DOI: 10.1016/j.biotechadv.2014.03.009

[127] Hai R, He L, Shu G, Yin G. Characterization of histone deacetylase mechanisms in cancer development. Frontiers in Oncology. 2021;**11**:700947

[128] Carlson SM, Gozani O. Nonhistone lysine methylation in the regulation of cancer pathways. Cold Spring Harbor Perspectives in Medicine. 2016;**6**(11):a026435

[129] Neganova ME, Klochkov SG, Aleksandrova YR, Aliev G. Histone

modifications in epigenetic regulation of cancer: Perspectives and achieved progress. In: Presented at the Seminars in Cancer Biology. Elsevier; 2022. pp. 452-471

[130] Pop S, Enciu AM, Tarcomnicu I, Gille E, Tanase C. Phytochemicals in cancer prevention: Modulating epigenetic alterations of DNA methylation. Phytochemistry Reviews. 2019;**18**:1005-1024

[131] Kang H. MicroRNA-mediated health-promoting effects of phytochemicals. International Journal of Molecular Sciences. 2019;**20**(10):2535

[132] Mohapatra P, Singh P, Singh D, Sahoo S, Sahoo SK. Phytochemical based nanomedicine: A panacea for cancer treatment, present status and future prospective. OpenNano. 2022;7:100055. DOI: 10.1016/j.onano.2022.100055

[133] Souho T, Lamboni L, Xiao L, Yang G. Cancer hallmarks and malignancy features: Gateway for improved targeted drug delivery. Biotechnology Advances. 2018;**36**(7):1928-1945. DOI: 10.1016/j.biotechadv.2018.08.001

[134] Efferth T. From ancient herb to modern drug: Artemisia annua and artemisinin for cancer therapy. Seminars in Cancer Biology. 2017;**46**:65-83. DOI: 10.1016/j.semcancer.2017.02.009

[135] Xie J, Yang Z, Zhou C, Zhu J, Lee RJ, Teng L. Nanotechnology for the delivery of phytochemicals in cancer therapy. Biotechnology Advances. 2016;**34**(4):343-353. DOI: 10.1016/j.biotechadv.2016.04.002

[136] Brys AK, Gowda R, Loriaux DB, Robertson GP, Mosca PJ. Nanotechnology-based strategies for combating toxicity and resistance in melanoma therapy. Biotechnology Advances. 2016;**34**(5):565-577. DOI: 10.1016/j.biotechadv.2016.01.004

[137] Santos IS, Ponte BM, Boonme P, Silva AM, Souto EB. Nanoencapsulation of polyphenols for protective effect against colon–rectal cancer. Biotechnology Advances. 2013;**31**(5):514-523. DOI: 10.1016/j.biotechadv.2012.08.005

[138] Chuan L et al. Biocompatible and biodegradable nanoparticles for enhancement of anti-cancer activities of phytochemicals. Chinese Journal of Natural Medicines. 2015;**13**(9):641-652

[139] Tsouris V, Joo MK, Kim SH, Kwon IC, Won Y-Y. Nano carriers that enable co-delivery of chemotherapy and RNAi agents for treatment of drug-resistant cancers. Biotechnology Advances. 2014;**32**(5):1037-1050. DOI: 10.1016/j.biotechadv.2014.05.006

[140] Kumar G, Virmani T, Sharma A, Pathak K. Codelivery of phytochemicals with conventional anticancer drugs in form of Nanocarriers. Pharmaceutics. 2023;**15**(3):Art. no. 3. DOI: 10.3390/pharmaceutics15030889

[141] Gupta V et al. Protein PEGylation for cancer therapy: Bench to bedside. Journal of Cell Communication and Signaling. 2019;**13**:319-330

[142] Rahman MA et al. Phytochemicals as a complement to cancer chemotherapy: Pharmacological modulation of the autophagy-apoptosis pathway. Frontiers in Pharmacology. 2021;**12**:639628. DOI: 10.3389/fphar.2021.639628

[143] Koklesova L et al. Phytochemical-based nanodrugs going beyond the state-of-the-art in cancer management—Targeting cancer stem cells in the

framework of predictive, preventive, personalized medicine. Frontiers in Pharmacology. 2023;**14**:1121950. DOI: 10.3389/fphar.2023.1121950

[144] Enrico C. Nanotechnology-based drug delivery of natural compounds and phytochemicals for the treatment of cancer and other diseases. Studies in Natural Products Chemistry. 2019;**62**:91-123

[145] Chavda VP et al. Advanced phytochemical-based nanocarrier systems for the treatment of breast cancer. Cancers. 2023;**15**(4):Art. no. 4. DOI: 10.3390/cancers15041023

**Chapter 8**

# Phytochemicals: Potential Drug Candidates for Leishmaniasis

*Nergiz Gürbüz Çolak*


## Abstract

Leishmaniasis is a neglected tropical disease with three main clinical manifestations. This disease is caused by protozoan parasites that are transmitted through the bite of female sandflies. Each year, more than one million cases of leishmaniasis are reported, and over one billion people are at risk of contacting the disease. Leishmaniasis is a complex disease that relapses when the immune system is suppressed. The currently available antileishmanial drugs have serious side effects. Therefore, it is important to identify novel drug candidates for leishmaniasis treatment. Phytochemicals are promising drug candidates because of their wide chemical diversity and bioavailability. This chapter provides an overview of leishmaniasis and highlights the phytochemicals that can be used to treat the disease.


**Keywords:** phytochemicals, bioactivity, drug candidates, neglected tropical diseases, leishmaniasis

## 1. Introduction

Among the numerous infectious diseases that affect humans, a significant majority, amounting to 60%, has an origin that is zoonotic in nature. These diseases not only have a detrimental impact on human and animal health but also on ecosystems and the economy [1]. Zoonotic diseases particularly affect vulnerable populations, such as women and children, in low- and middle-income regions. These diseases are often referred to as "neglected" diseases, a term that reflects the limited funding available for their prevention and control [2]. The World Health Organization (WHO) recently developed a roadmap to combat 20 neglected tropical diseases, including leishmaniasis [3]. This roadmap outlines the strategies for controlling and eliminating leishmaniasis, including the need for new treatment approaches.

Development of new drugs to combat leishmaniasis, particularly in endemic countries, is of utmost importance. The treatment of leishmaniasis is heavily dependent on a limited number of medications such as pentavalent antimonials, amphotericin B, miltefosine, pentamidine, and paromomycin. Nevertheless, these drugs have several limitations, including increased treatment failure rates, high costs, and significant side effects, which collectively compromise their effectiveness [4]. Secondary plant metabolism plays a pivotal role in the discovery of novel medications against protozoal infections, including chemotherapeutic agents. This could potentially lead to the identification of novel and selective antileishmanial compounds [5].

## 2. Overview of *Leishmania* and Leishmaniasis

Leishmaniasis is a disease caused by the *Leishmania* parasite, a eukaryotic protozoan belonging to the class Trypansomatidae and order Kinetoplastida [6]. Over 20 species of *Leishmania* are known to be pathogenic to humans [7]. The transmission of this disease is facilitated by the bites of female phlebotomine sandflies, more specifically *Phlebotomus* in the Old World [8] and *Lutzomyia* in the New World [9]. According to the Pan American Health Organization, leishmaniasis poses a significant threat to global health, as it is prevalent in 99 countries [10]. Each year, this disease is estimated to cause approximately one million new cases of cutaneous leishmaniasis and 30,000 new cases of visceral leishmaniasis, as reported by the World Health Organization [11]. The economic consequences of the disease are severe due to extended treatment regimens and hospitalization, which could be fatal if left untreated [12]. Every year, 20,000–30,000 deaths occur worldwide [8]. It is crucial to recognize that the number of reported cases represents only a fraction of the actual cases, as individuals with asymptomatic infections are typically unaware of their condition [13]. Consequently, the actual number of cases is likely greater than the reported number.

### 2.1 Mode of transmission

Transmission of the *Leishmania* parasite mainly occurs via vectors, specifically through the bite of sand flies, but it can also spread in other ways. For instance, transmission can occur through blood transfusions from asymptomatic individuals who are infected, sharing of needles among drug users, transmission via the placenta to the fetus in the womb, transmission through sexual contact, and contamination of laboratory samples [14].

### 2.2 Life cycle of *Leishmania*

*Leishmania* exists in two primary forms, promastigotes and amastigotes, which differ morphologically. Promastigotes are extracellular, elongated, and flagellated, whereas amastigotes are intracellular and spherical [15]. The parasite's life cycle alternates between host and carrier, undergoing morphological changes at each stage [16]. Female sand flies acquire parasites by ingesting blood from infected hosts. Inside the sandfly, the parasite develops into a procyclic promastigote that attaches to the gut lining using flagella [17]. Procyclic forms are then transformed into nectomonad promastigotes in the abdominal midgut. These promastigotes migrate to the anterior midgut, become leptomonads, and differentiate into metacyclic promastigotes. Metacyclic promastigotes travel to insect proboscises for transmission to mammalian hosts. Metacyclogenesis involves the introduction of promastigotes into the host via sandfly bite, initiating a phagocytic process in which promastigotes enter phagocytes and attack parasitophorous vacuoles. Inside the vacuole, promastigotes transform into oval-shaped amastigotes, which multiply until macrophages rupture, thereby releasing mature amastigotes. This release triggers a chain reaction that leads to leishmaniasis [14, 18].

### 2.3 Clinical forms of Leishmaniasis

Leishmaniasis can take three primary forms, cutaneous (CL), mucocutaneous (MCL), and visceral leishmaniasis (VCL), which are determined by the specific clinical features of the infection and the type of parasite causing it [19]. The clinical presentation of the disease is influenced by the intricate relationship between the parasite and the host's immune response. It can range from minor skin lesions to severe, potentially life-threatening illnesses [20]. Cutaneous leishmaniasis (CL), one of the most common form, is caused by various *Leishmania* species, including *L. major, L. aethiopica, L. mexicana, L. tropica, L. braziliensis, L. venezuelensis, L. peruviana, L. amazonensis, L. panamensis*, and *L. guyanensis*. CL typically manifests as painful ulcerative nodular lesions on the hands, face, and feet that may result in scarring. Diffuse cutaneous leishmaniasis (DCL), a form of CL, is a widely distributed non-ulcerative skin lesion primarily caused by *L. aethiopica* and *L. amazonensis* [14]. Mucocutaneous leishmaniasis (MCL) involves progressive destruction of the nasal mucosa and pharyngeal regions, leading to difficulties in food consumption and an increased risk of secondary infections. However, *L. braziliensis, L. amazonensis, L. mexicana, L. panamensis*, and *L. aethiopica* are the primary causative species, *L. tropica, L. infantum, and L. major* have also been reported [14]. Visceral leishmaniasis (VL), or Kala-azar, is the most severe form, primarily caused by *L. chagasi, L. donovani, L. tropica,* and *L. infantum*. Symptoms range from asymptomatic infections to life-threatening conditions, with a significant risk of secondary infections that complicate treatment. In addition, VL can be fatal if left untreated. Moreover, even if treated, it may progress to post-Kala-azar dermal leishmaniasis, which can cause extensive damage to the mucous membranes of the nose, throat, and oral cavity, resulting in disfiguring lesions [21].

### 2.4 Current treatment options and limitations

Leishmaniasis treatment depends on the *Leishmania* species, patient's clinical condition, medication availability, and possible side effects. Common treatments include pentavalent antimonials, amphotericin B, miltefosine, pentamidine, and paramomycin, each with distinct advantages and disadvantages [14].

*Antimonial compounds*, pentavalent antimony (SbV), are known as first-line therapies and have been used to treat leishmaniasis since 1945. Commercially available formulations of antimonials are Glucantime® and Pentostam® [22], and their efficacy varies for visceral and cutaneous leishmaniasis; however, they remain the most common treatments [23]. The main reasons for the limited application of antimonials are their adverse reactions/toxicity and emergence of resistance and treatment failures in countries within the subcontinent. Patients often report local pain from intramuscular injections and severe side effects, such as nausea, vomiting, appetite loss, muscle and abdominal pain, headache, joint pain, and fatigue [24]. In rare cases, these medications may cause fatal cardiac arrhythmias [25].

*Amphotericin B (AmB)* is a broad-spectrum antifungal used to treat leishmaniasis, including the initial treatment of visceral leishmaniasis (VL) in India due to antimonial drug ineffectiveness [26]. It is effective against multiple *Leishmania* species and is recommended for pregnant women and individuals coinfected with Leishmania and HIV. While the efficacy of AmB exceeds 90%, it is associated with significant toxicity and high cost [27]. Patients may experience adverse reactions that require hospital observation, such as fever, myocarditis, and nephrotoxicity [28]. Lipid-based formulations, such as AmBisome®, Amphocil®, and Abelcet®, have been developed to reduce these adverse effects [29]. Liposomal amphotericin B has fewer adverse effects and is preferred as a first-line treatment in Brazil [30]. Liposomal AmB, which is available at lower costs (Fungisome®), and other delivery systems offer feasible solutions for low-income regions [31].

*Miltefosine*, an alkyl phosphocholine derivative, was initially identified as an antineoplastic agent for cutaneous neoplasm [32]. In the 1980s, it was found to have antileishmanial properties and has since become a widely accepted chemotherapy option for leishmaniasis, showing clinical effectiveness comparable to AmB [33]. Its primary advantage is that it is the only approved oral medication for leishmaniasis [34]. However, the long half-life (> 120 h) and teratogenic potential of miltefosine restrict its use during pregnancy. Additionally, its limited efficacy in certain leishmaniasis types, necessity for high doses, and emerging resistance are significant concerns [24].

*Pentamidine* serves as a second-line treatment for leishmaniasis by enhancing the immune system's ability to fight the parasite [35]. It is administered intravenously or intramuscularly to patients who are unresponsive to pentavalent antimony [36]. Despite its efficacy, pentamidine exhibits high toxicity, including cardiotoxicity, hypotension, and irreversible insulin-dependent diabetes mellitus [37], particularly in visceral leishmaniasis. Additionally, resistance has been reported to result from increased drug efflux [38].

*Paromomycin*, an aminoglycoside antibiotic, has demonstrated efficacy in combating leishmanial infections [39]. This drug is utilized in certain areas of the Indian subcontinent for both systemic and topical purposes in the treatment of VL, as well as for the systemic treatment of CL [40]. Despite its low cost and lack of significant toxicity, the use of paromomycin as a monotherapy is hindered by the emergence of resistance [24].

Current therapeutic strategies for leishmaniasis face considerable obstacles, including high cost, toxicity, emergence of drug resistance, various routes of administration, extended treatment periods, and documented treatment failures. The rising issue of severe resistance to medications such as amphotericin B and miltefosine, especially in regions with recurrent outbreaks, is a matter of great concern [24]. Furthermore, the control and treatment of leishmaniasis are complicated by factors such as lack of public awareness and knowledge, low socioeconomic conditions, malnutrition, immunosuppressive medication use, weak healthcare systems, misdiagnosis, delayed or undiagnosed patients, climate change, unfeasibility of vector and reservoir control, parasite drug resistance, and absence of a vaccine [41]. These factors also compromise the effectiveness of the current treatment approaches. Therefore, there is an urgent need to discover and develop new drug candidates for leishmaniasis treatment.

## 3. Phytochemicals as a potential source for Leishmaniasis

The contemporary therapeutic choices for leishmaniasis are becoming less effective as a result of the emergence of resistance, which poses a significant challenge [42]. Many natural products possess exceptional biological properties, making them promising sources of new compounds for use in drug development. These products can serve as the basis for the creation of new agents to treat a range of diseases, including neglected tropical diseases such as leishmaniasis. The use of plants for medicinal purposes has a history dating back thousands of years [43]. Plants have traditionally been employed in unprocessed forms, such as teas, powders, tinctures, poultices, and other herbal preparations. During the early ninteenth century, the use of plants as medicines progressed to involve the isolation of active compounds, starting with the extraction of morphine from the opium [43]. Several notable active compounds, including betulinic acid, combretastatin A4 phosphate, harpagoside [44], quinine, artemisinin [45], digitoxin, paclitaxel, and pilocarpin [46], have been extracted from plants and are currently being used as drugs. In numerous tropical countries, conventional leishmaniasis therapy commonly entails oral administration of crude plant extracts for the systemic form of the disease, whereas topical treatments are applied to treat skin infections [47].

Medicinal plants possess considerable potential for the production of compounds capable of effectively treating a wide variety of ailments owing to their substantial concentration of potent metabolites. Therefore, researchers are increasingly focusing on natural, sustainable sources for the effective management of parasitic diseases and their complications. This approach offers several advantages, including fewer side effects, lower costs, and environmental friendliness [48]. In addition, the pharmaceutical sector has recently shifted its attention towards plant-based medicines and formulations, as these tend to exhibit low toxicity and heightened efficacy [42]. The untapped potential of plants as sources of new medications is yet to be fully harnessed. Despite the fact that there are approximately 250,000 medicinal plants, only a small percentage of them have had their biological activities studied, with only 6% being known, and a mere 0.75% of herbal compounds has been scientifically investigated [49].

### 3.1 Limitations of phytochemicals

The use of plant-based therapies offers several advantages, including reduced likelihood of adverse reactions and high potency [50]. However, most of these compounds have only demonstrated effectiveness in *in vitro* studies and have not advanced to the pre-clinical trial stage [14]. In addition, the pharmaceutical industry faces several obstacles when attempting to commercialize natural products. One significant challenge is to precisely identify and assign taxonomic names to plant sources that are frequently harvested from their natural habitats. The scarcity of these natural products further complicates the assessment of their broad range of biological activities. The chemical makeup of plant materials is critical and is influenced by various factors such as species type, harvest timing, soil composition, altitude, climate conditions, and storage practices. Moreover, the extraction and isolation processes can lead to the degradation of compounds. Addressing the challenges posed by incompatibility between crude extracts and high-throughput assay procedures in natural product research is crucial. One of the major challenges in this field is understanding the precise molecular mechanisms of plant extracts, which are compounded by inconsistent experimental results and the high costs associated with obtaining natural product samples. The presence of artifacts in some extracts also hinders progress, making it difficult to isolate the active compounds from these extracts. Obtaining substantial quantities of these substances is particularly challenging because of their natural origin and lengthy resupply times. The slow growth and sparse distribution of certain species exacerbate these difficulties. Moreover, compliance with the Convention on Biological Diversity poses additional obstacles, especially when a plant is commercialized for use as a herbal remedy or in pharmaceutical drug production, leading to risks from extensive wildcrafting and unsustainable harvesting practices [51, 52].

### 3.2 Antileishmanial phytochemicals: Literature examples

An isolated compound, luteoside B, from stem bark of *Tecoma mollis* [53], an isolated labdane diterpenoid, sceptrumlabdalactone B, from rhizomes of *Aframomum spectrum* K. Schum [54], methanolic extract of stem bark of *Sterculia villosa* [55], ethyl acetate and dichlormethane extracts of husks of *Acacia nilotica,* ethyl acetate, dichlormethane, pethrolium ether and water extracts of aerial shoots of *Ambrosia miratima* [56], an isoquinoline alkaloid, (−) caaverine, isolated from stem bark of *Ocotea lancifolia* [57], semi-purified fractions of ethyl acetate extract of leaves of *Parthenium hysterophorus* [58], an isolated compund, emodin, from rhizomes of *Rumex abyssinicus* [59], an isolated compound, putranoside A, from leaves of *Drypetes gerrardii* [60], hydroethanolic extract of aerial parts of *Cicer microphyllum* [61], crude methanolic extract and isolated compunds, funtumic acid A and B, from leaves of *Funtumia*

| ***L. donovani*** **promatsigotes** | | | |
|---|---|---|---|
| **Plant** | **Plant part** | **Extract/isolated compound(s)** | **Reference** |
| *Tecoma mollis* | Stem bark | Luteoside B | [53] |
| *Aframomum spectrum* K. Schum | Rhizomes | Sceptrumlabdalactone B | [54] |
| *Sterculia villosa* | Stem bark | Methanolic extract | [55] |
| *Acacia nilotica* | Husks | Ethyl acetate and dichlormethane extracts | [56] |
| *Ambrosia miratima* | Aerial shoots | Ethyl acetate, dichlormethane, pethrolium ether, and water extracts | [56] |
| *Ocotea lancifolia* | Stem bark | (−) Caaverine | [57] |
| *Parthenium hysterophorus* | Leaves | Semi-purified fractions of ethyl acetate extract | [58] |
| *Rumex abyssinicus* | Rhizomes | Emodin | [59] |
| *Drypetes gerrardii* | Leaves | Putranoside A | [60] |
| *Cicer microphyllum* | Aerial parts | Hydroethanolic extract | [61] |
| *Funtumia elastica* | Leaves | Crude methanolic extract | [62] |
| *Funtumia elastica* | Leaves | Funtumic acid A and B | [62] |
| *Funtumia elastica* Preuss Stapf. | Trunk bark | Funtulatine | [63] |
| *Delphinium gracile* | N/A | Newly identified | [64] |
| *Cycas revoluta* | Leaves | Vitexin-2″-rhamnoside | [65] |
| *Lantana montevidensis* | Roots | Oleanonic acid | [66] |
| *Dodonaea viscosa* Linn. | Aerial parts | 5-Hydroxy-3,6,7,4′ -tetramethoxy flavone and aliarin | [67] |
| *Quetzalia ilicina* | Root bark | Acetone extract | [68] |
| *Garcinia cowa* | Leaves and barks | Methanolic extract | [69] |
| *Lantana ukambensis* | N/A | Methanolic extract | [70] |
| *Prosopis juliflora* | Leaves | Methanolic extract | [71] |
| *Clematis simensis fresen* | Leaves | Hydromethanolic extract | [72] |
| *Euphorbia abyssinica* | Latex | Hydromethanolic extract | [72] |
| *Calea zacatechichi* Schl. | Aerial parts | Germacranolides | [73] |

**Table 1.**
*Antileishmanial activity of extracts and isolated compounds against* L. donovani *promatsigotes.*

*elastica* [62], an isolated compound, funtulatine, from trunk bark of *Funtumia elastica* Preuss Stapf. [63], newly isolated compunds from *Delphinium gracile* [64], an isolated compound, vitexin-2″-rhamnoside, from leaves of *Cycas revoluta* [65], an isolated compound, oleanonic acid, from roots of *Lantana montevidensis* [66], isolated compounds, 5-hydroxy-3,6,7,4′-tetramethoxy flavone and aliarin, from aerial parts of *Dodonaea viscosa* Linn. [67], acetone extract of root bark of *Quetzalia ilicina* [68], methanolic extract of leaves and barks of *Garcinia cowa* [69], methanolic extract of *Lantana ukambensis* [70], methanolic extract of leaves of *Prosopis juliflora* [71], hydro-methanolic extract of leaves of *Clematis simensis fresen* and latex of *Euphorbia abyssinica* [72], and isolated germacranolides from aerial parts of *Calea zacatechichi* Schl. [73] exhibited antileishmanial activity against *L. donovani* promatsigotes (**Table 1**).

A newly isolated compound from leaves of *Piliostigma thonningii* [74], dichloromethane fraction of methanol extract of aerial parts of *Helichrysum oligocephal* [75], isolated napththylisoquinoline alkaloids, ancistrogriffine A and C, from leaves of *Ancistrocladus griffithii* [76], and ancistrotanzanine A and B, from leaves of *Ancistrocladus tanzaniensis* [77], an isolated abietane diterpenoid, leriifoliol, from roots of *Salvia leriifolia* [78], an isoflavanol, pumilanol, isolated from roots of *Tephrosia pumila* [79], an isolated compound, emodin, from rhizomes of *Rumex abyssinicus* [59], isolated compounds, 7-deacetylgedunin and 7-deacetyl-7-oxogedunin, from roots of *Pseudocedrela kotschyi* [80], isolated compounds, apigenin and luteolin, from aerial parts of *Melampyrum arvense* [81], isolated biflavanoids, 2,3-dihydrohinokiflavone and 7-O-methylhinokiflavone, from *Selaginella bryopteris* [82], an isolated compound, geranylfarnesol, from aerial parts of *Thalia geniculata* [83], a diarylheptanoid, 1,7-bis-(4-hydroxyphenyl)-3-hepten-5-one, from aerial parts of *Schrankia leptocarpa* DC. [84], an isolated compound, guttiferone A, from fruits of *Allanblackia monticola* and isolated compound, guttiferone F, from leaves of *Symphonia globulifera* [85], isolated compounds, betunilic acid and quarcetrin, from leaves of *Pelliciera rhizophorae* [86], newly isolated compunds from *Delphinium gracile* [64], isolated compounds, vismione D and emodin, from stem bark of *Vismia orientalis* Engl. [87], an isolated compound, oleanonic acid, from roots of *Lantana montevidensis* [66], aqueous extracts of aerial parts of *Xanthium brasilicum* Vell., and isolated compounds, 4,15-dinor-1,11(13)-xanthadiene-3,5β:12,8β-diolide and 8-epixanthatin 1β,5β-epoxide, from aerial parts of Asteraceae species native to Sudan [88], dichloromethane extract of aerial parts of *Ageratum conyzoides* L. [89], dichloromethane extract of aerial parts of *Larrea tridentata* [90], an isolated compound, 2-acetylnaphtho[2,3-b]furan-4,9-dione, from stem bark of *Markhamia tomentosa* [91], dichloromethane extract of leaves of *Piper casapiense*, *Piper acutilimbum*, *Piper lineatum*, and methanolic extract of leaves of *Piper pseudoarboreum* [92], hexane extract of roots of *Zhumeria majdae* [93] exhibited antileishmanial activity against *L. donovani* amastigotes (**Table 2**).

Methanolic extract of *Teucrium oliverianum* [94], methanolic extract and isolated compound, hesperidin, from flowers of *Calendula officinalis* L. [45], isolated physalins from *Physalis minima* [42], chloroform and ethyl acetate extracts from stems of *Tocoyena hispidula* [95], isolated steroidal alkaloids, hookerianamide J, K, H and I, vagenine A,2,3-dehydrosarsalignone and sarcovagine C, from *Sarcococca hookeriana* [96], an isolated sesquiterpene lactone, 1b,2b,8a-trihydroxy-11b,13-dihydroelema-12,6a-olide, from aerial parts of *Seriphidium khorassanicum* [97], ethyl acetate extract of aerial parts of *Echium arenarium* (Guss) [98], chloroform extracts of *Euphorbia retusa* and *Pulicaria undulata* [99], isolated compounds, oleanolic acid and ursolic acid, from leaves of *Tabebuia aurea* [100], isolated compounds,

5,3′,4′-trihydroxy-6-(g,g-dimethylallyl)-6″,6″-dimethylpyrano-(2″,3″,7,8)-flavanone, dorsmanine I, lonchocarpol A and 6,8-diprenyleriodictyol, from stem and leaves of *Derris ferruginea* [101], an isolated steroidal saponin from bulbs of *Allium paradoxum* L. [102], isolated compounds, minutin A and B, 8-hydroxyisocapnolactone-2′,3′ -diol and clauslactone E, from leaves of *Micromelum minutum* [103], an isolated compound,

| ***L. donovani*** **amastigotes** | | | |
|---|---|---|---|
| **Plant** | **Plant part** | **Extract/isolated compound(s)** | **Reference** |
| *Rumex abyssinicus* | Rhizomes | Emodin | [59] |
| *Delphinium gracile* | N/A | Newly identified | [64] |
| *Lantana montevidensis* | Roots | Oleanonic acid | [66] |
| *Piliostigma thonningii* | Leaves | Newly identified | [74] |
| *Helichrysum oligocephal* | Aerial parts | Dichloromethane fraction of methanol extract | [75] |
| *Ancistrocladus griffithii* | Leaves | Ancistrogriffine A and C | [76] |
| *Ancistrocladus tanzaniensis* | Leaves | Ancistrotanzanine A and B | [77] |
| *Salvia leriifolia* | Roots | Leriifoliol | [78] |
| *Tephrosia pumila* | Roots | Pumilanol | [79] |
| *Pseudocedrela kotschyi* | Roots | 7-Deacetylgedunin and 7-deacetyl-7-oxogedunin | [80] |
| *Melampyrum arvense* | Aerial parts | Apigenin and luteolin | [81] |
| Selaginella bryopteris | N/A | 2,3-Dihydrohinokiflavone and 7-O-methylhinokiflavone | [82] |
| Thalia geniculata | Aerial parts | Geranylfarnesol | [83] |
| *Schrankia leptocarpa* DC. | Aerial parts | 1,7-Bis-(4-hydroxyphenyl)-3-hepten-5-one | [84] |
| *Allanblackia monticola* | Fruits | Guttiferone A | [85] |
| *Symphonia globulifera* | Leaves | Guttiferone F | [85] |
| *Pelliciera rhizophorae* | Leaves | Betunilic acid and quarcetrin | [86] |
| *Vismia orientalis* Engl. | Stem bark | Vismione D and emodin | [87] |
| *Xanthium brasilicum* Vell. | Aerial parts | Aqueous extracts | [88] |
| *Asteraceae* species native to Sudan | Aerial parts | 4,15-Dinor-1,11(13)-xanthadiene-3,5β:12,8β-diolide and 8-epixanthatin 1β,5β-epoxide | [88] |
| *Ageratum conyzoides* L. | Aerial parts | Dichloromethane extract | [89] |
| *Larrea tridentata* | Aerial parts | Dichloromethane extract | [90] |
| Markhamia tomentosa | Stem bark | 2-Acetylnaphtho[2,3-b]furan-4,9-dione | [91] |
| *Piper casapiense* | Leaves | Dichloromethane extract | [92] |
| *Piper acutilimbum* | Leaves | Dichloromethane extract | [92] |
| *Piper lineatum* | Leaves | Dichloromethane extract | [92] |
| *Piper pseudoarboreum* | Leaves | Methanolic extract | [92] |
| *Zhumeria majdae* | Roots | Hexane extract | [93] |

**Table 2.**
*Antileishmanial activity of extracts and isolated compounds against* L. donovani *amastigotes.*

chamuangone, from leaves of *Garcinia cowa* [104], methanolic extract of *Veronica persica* Poir. [105], chloroform extract of *Gossypium hirsutum* [106], methanolic extract of stem bark of *Azadirchta indica* [107], an isolated triterpene, pistagremic acid, from galls of *Pistacia integerrima* Stewart [108], isolated compounds, crossoseamine A and B [109], and crossostenin A, B, C and D, and crossoseamine C from aerial parts of *Crossostephium chinense* [110] exhibited antileishmanial activity against *L. major* promatsigotes (**Table 3**).

Methanolic extract of *Teucrium oliverianum, Pergularia tomentosa* and *Cleome amblyocarpa* [94], isolated compounds, hesperidin and rutin, from flowers of

| ***L. major* promatsigotes** | | | |
|---|---|---|---|
| **Plant** | **Plant part** | **Extract/isolated compound(s)** | **Reference** |
| *Physalis minima* | N/A | Physalins | [42] |
| *Calendula officinalis* L. | Flowers | Methanolic extract | [45] |
| *Calendula officinalis* L. | Flowers | Hesperidin | [45] |
| *Teucrium oliverianum* | N/A | Methanolic extract | [94] |
| *Tocoyena hispidula* | Stems | Chloroform and ethyl acetate extracts | [95] |
| *Sarcococca hookeriana* | N/A | Hookerianamide J, K, H, and I, vagenine A,2,3-dehydrosarsalignone and sarcovagine C | [96] |
| *Seriphidium khorassanicum* | Aerial parts | 1b,2b,8a-Trihydroxy-11b,13-dihydroelema-12,6a-olide | [97] |
| *Echium arenarium* (Guss) | Aerial parts | Ethyl acetate extract | [98] |
| *Euphorbia retusa* | N/A | Chloroform extract | [99] |
| *Pulicaria undulata* | N/A | Chloroform extract | [99] |
| *Tabebuia aurea* | Leaves | Oleanolic acid and ursolic acid | [100] |
| *Derris ferruginea* | Stems and leaves | 5,3′,4′-Trihydroxy-6-(g,g-dimethylallyl)-6″,6″-dimethylpyrano-(2″,3″:7,8)-flavanone, dorsmanine I, lonchocarpol A and 6,8-diprenyleriodictyol | [101] |
| *Allium paradoxum* L. | Bulbs | Steroidal saponin | [102] |
| *Micromelum minutum* | Leaves | Minutin A and B, 8-hydroxyisocapnolactone-2′,3′-diol and clauslactone E | [103] |
| *Garcinia cowa* | Leaves | Chamuangone | [104] |
| *Veronica persica* Poir. | N/A | Methanolic extract | [105] |
| *Gossypium hirsutum* | N/A | Chloroform extract | [106] |
| *Azadirchta indica* | Stem bark | Methanolic extract | [107] |
| *Pistacia integerrima* Stewart | Galls | Pistagremic acid | [108] |
| *Crossostephium chinense* | Aerial parts | Crossoseamine A and B | [109] |
| *Crossostephium chinense* | Aerial parts | Crossostenin A, B, C, and D and crossoseamine C | [110] |

**Table 3.**
*Antileishmanial activity of extracts and isolated compounds against* L. major *promatsigotes.*

| ***L. major* amastigotes** | | | |
|---|---|---|---|
| **Plant** | **Plant part** | **Extract/isolated compound(s)** | **Reference** |
| *Calendula officinalis* L. | Flowers | Hesperidin and rutin | [45] |
| *Teucrium oliverianum* | N/A | Methanolic extract | [94] |
| *Pergularia tomentosa* | N/A | Methanolic extract | [94] |
| *Cleome amblyocarpa* | N/A | Methanolic extract | [94] |
| *Euphorbia retusa* | N/A | Chloroform extract | [99] |
| *Pulicaria undulata* | N/A | Chloroform extract | [99] |
| *Ziziphus spina-christi* | Aerial parts | Methanolic extract | [111] |
| *Curcuma longa* Linn. | N/A | Curcumin | [112] |

**Table 4.**
*Antileishmanial activity of extracts and isolated compounds against* L. major *amastigotes.*

*Calendula officinalis* L. [45], methanolic extract of aerial parts of *Ziziphus spina-christi* [111], isolated compound, curcumin, from *Curcuma longa* Linn [112], and chloroform extracts of *Euphorbia retusa* and *Pulicaria undulata* [99] exhibited antileishmanial activity against *L. major* amastigotes (**Table 4**).

Ethanolic extract of leaves of *Senna rugosa* [113], isolated compounds, 3-chlorosintenpyridone and (S)-13-hydroxy-octadeca-(9Z,11E,15Z)-trienoic acid, from leaves of *Piper pseudoarboreum* [114], ethyl acetate extract of aerial parts of *Echium arenarium* (Guss) [98], newly isolated compunds from *Delphinium gracile* [64], semisynthetic terpenoid derivatives from latex of *Euphorbia resinifera* and *Euphorbia officinarum* [115], and dichloromethane and methanol fraction of radix of *Pycnarrhena cauliflora* [116] exhibited antileishmanial activity against *L. infantum* promatsigotes (**Table 5**).

Dichloromethane extract and isolated compound, α-mangostin, from *Garcinia mangostana* [117], essential oil of leaves of *Piper angustifoliu* [118], neem seed oil of *Azadirachta indica* A. Juss [119], ethyl acetate extract of aerial parts of *Echium arenarium* (Guss) [98], newly isolated compounds from *Delphinium gracile* [64], and methanolic extract of aerial parts of *Hypoestes forskalei* [120] exhibited antileishmanial activity against *L. infantum* amastigotes (**Table 6**).

| ***L. infantum* promatsigotes** | | | |
|---|---|---|---|
| **Plant** | **Plant part** | **Extract/isolated compound(s)** | **Reference** |
| *Delphinium gracile* | N/A | Newly identified | [64] |
| *Echium arenarium* (Guss) | Aerial parts | Ethyl acetate extract | [98] |
| *Senna rugosa* | Leaves | Ethanolic extract | [113] |
| *Piper pseudoarboreum* | Leaves | 3-Chlorosintenpyridone and (S)-13-hydroxy-octadeca-(9Z,11E,15Z)-trienoic acid | [114] |
| *Euphorbia resinifera* | Latex | Semisynthetic terpenoid derivatives | [115] |
| *Euphorbia officinarum* | Latex | Semisynthetic terpenoid derivatives | [115] |
| *Pycnarrhena cauliflora* | Radix | Dichloromethane and methanol fraction | [116] |

**Table 5.**
*Antileishmanial activity of extracts and isolated compounds against* L. infantum *promatsigotes.*

| ***L. infantum* amastigotes** | | | |
|---|---|---|---|
| **Plant** | **Plant part** | **Extract/isolated compound(s)** | **Reference** |
| *Delphinium gracile* | N/A | Newly identified | [64] |
| *Echium arenarium* (Guss) | Aerial parts | Ethyl acetate extract | [98] |
| *Garcinia mangostana* | N/A | Dichloromethane extract | [117] |
| *Garcinia mangostana* | N/A | α-Mangostin | [117] |
| *Piper angustifoliu* | Leaves | Essential oil | [118] |
| *Azadirachta indica* A. Juss | Seeds | Neem seed oil | [119] |
| *Hypoestes forskalei* | Aerial parts | Methanolic extract | [120] |

**Table 6.**
*Antileishmanial activity of extracts and isolated compounds against* L. infantum *amastigotes.*

An isolated compound, lupenone, from aerial parts of *Albizia inundata* [121], methanolic extract of fruits of *Libidibia ferrea* [122], ethanolic extract of leaves of *Senna rugosa* [113], essential oils of *Copaifera reticulate* (Para), *Copaifera reticulata* (Acre), *Copaifera martii, Copaifera cearensis, Copaifera paupera, Copaifera langsdorffii, Copaifera officinalis, Copaifera multijuga* and *Copaifera lucens* [123], isolated compounds, kaurenoic acid, hydroxycopalic acid, polyalthic acid and pinifolic acid, from copaiba oil [124], aqueous extract of root bark of *Juglans regia* L. [125], hydromethanolic extract of aerial parts of *Mitracarpus frigidus* [126], volatile oil of leaves of *Psidium firmum* O. BERG [127], isolated compounds, 3-chlorosintenpyridone and (S)-13-hydroxy-octadeca-(9Z,11E,15Z)-trienoic acid, from leaves of *Piper pseudoarboreum* [114], an isoquinoline alkaloid, (−) caaverine, isolated from stem bark of *Ocotea lancifolia* [57], an isolated alkaloids, 14-oxoprunifoleine and strictosamide, from roots of *Psychotria prunifolia* [128], dichloromethane extract and an isolated oxoaporphine alkaloid, liriodenine, from leaves of *Annona mucosa* [129], isolated compound, trixanolide, from aerial parts of *Trixis antimenorrhoea* [130], acetone extract of root bark of *Quetzalia ilicina* [68], dichloromethane fraction of hydroalcoholic extract of aerial parts of *Tanacetum parthenium* (L.) [131], and dichloromethane extract of flowers of *Chromolaeana hirsuta* [132] exhibited antileishmanial activity against *L. amazonensis* promatsigotes (**Table 7**).

An isolated compound, lupenone, from aerial parts of *Albizia inundata* [121], isolated xanthones from stem bark of *Allanblackia gabonensis* [133], isolated sesquiterpenes from leaves of *Renealmia thyrsoidea* [134], methanolic extract of fruits of *Libidibia ferrea* [122], ethyl acetate extracts of aerial parts of *Coronilla scorpioides, Eremanthus incanus* and *Pseudogynoxys cabrerae* [135], isolated compounds, kaurenoic acid, hydroxycopalic acid, polyalthic acid and pinifolic acid, from copaiba oil [124], hydroethanolic extracts of roots and leaves of *Leonotis nepetifolia* (L.) Br. [136], dichloromethane fraction of hydroalcoholic extract of aerial parts of *Tanacetum parthenium* (L.) [131], hydroethanolic extract and isolated biflavanoids, amentoflavone and robustaflavone, from *Selaginella sellowii* [137], and an isolated compound, 2′-hydroxy-3′,4′,6′-trimethoxychalcone, from leaves of *Piper hispidum* [138] exhibited antileishmanial activity against *L. amazonensis* amastigotes (**Table 8**).

Ethanolic extract of leaves of *Tarenaya spinosa* (Jacq.) Raf. [139], isolated compounds, deacetylspathelin and casimiroin, isolated compounds, 3-chlorosintenpyridone and (S)-13-hydroxy-octadeca-(9Z,11E,15Z)-trienoic acid, from leaves of *Piper pseudoarboreum* [114], an isoquinoline alkaloid, (−) caaverine, isolated from stem

| **L. *amazonensis* promatsigotes** | | | |
|---|---|---|---|
| **Plant** | **Plant part** | **Extract/isolated compound(s)** | **Reference** |
| *Ocotea lancifolia* | Stem bark | (−) Caaverine | [57] |
| *Quetzalia ilicina* | Root bark | Acetone extract | [68] |
| *Senna rugosa* | Leaves | Ethanolic extract | [113] |
| *Piper pseudoarboreum* | Leaves | 3-Chlorosintenpyridone and (S)-13-hydroxy-octadeca-(9Z,11E,15Z)-trienoic acid | [114] |
| *Albizia inundata* | Aerial parts | Lupenone | [121] |
| *Libidibia ferrea* | Fruits | Methanolic extract | [122] |
| *Copaifera reticulate* (Para) | N/A | Essential oil | [123] |
| *Copaifera reticulata* (Acre) | N/A | Essential oil | [123] |
| *Copaifera martii* | N/A | Essential oil | [123] |
| *Copaifera cearensis* | N/A | Essential oil | [123] |
| *Copaifera paupera* | N/A | Essential oil | [123] |
| *Copaifera langsdorffii* | N/A | Essential oil | [123] |
| *Copaifera officinalis* | N/A | Essential oil | [123] |
| *Copaifera multijuga* | N/A | Essential oil | [123] |
| *Copaifera lucens* | N/A | Essential oil | [123] |
| Copaiba oil | N/A | Kaurenoic acid, hydroxycopalic acid, polyalthic acid, and pinifolic acid | [124] |
| *Juglans regia* L. | Root bark | Aqueous extract | [125] |
| *Mitracarpus frigidus* | Aerial parts | Hydromethanolic extract | [126] |
| *Psidium firmum* O. BERG | Leaves | Volatile oil | [127] |
| *Psychotria prunifolia* | Roots | 14-Oxoprunifoleine and strictosamide | [128] |
| *Annona mucosa* | Leaves | Dichloromethane extract | [129] |
| *Annona mucosa* | Leaves | Liriodenine | [129] |
| *Trixis antimenorrhoea* | Aerial parts | Trixanolide | [130] |
| *Tanacetum parthenium* (L.) | Aerial parts | Dichloromethane fraction of hydroalcoholic extract | [131] |
| *Chromolaeana hirsuta* | Flowers | Dichloromethane extract | [132] |

**Table 7.**
*Antileishmanial activity of extracts and isolated compounds against* L. amazonensis *promatsigotes.*

bark of *Ocotea lancifolia* [57], dichloromethane extract and an isolated oxoaporphine alkaloid, liriodenine, from leaves of *Annona mucosa* [129], isolated compound, trixanolide, from aerial parts of *Trixis antimenorrhoea* [130], newly isolated compunds from *Delphinium gracile, Consolida oliveriana* and *Aconitum napellus* subsp. Lusitanicum [64], ethanol extract of leaves of *Pityrogramma calomelanos* (L.) [140],

| ***L. amazonensis* amastigotes** | | | |
|---|---|---|---|
| **Plant** | **Plant part** | **Extract/isolated compound(s)** | **Reference** |
| *Albizia inundata* | Aerial parts | Lupenone | [121] |
| *Libidibia ferrea* | Fruits | Methanolic extract | [122] |
| Copaiba oil | N/A | Kaurenoic acid, hydroxycopalic acid, polyalthic acid, and pinifolic acid | [124] |
| Allanblackia gabonensis | Stem bark | Xanthones | [133] |
| *Renealmia thyrsoidea* | Leaves | Sesquiterpenes | [134] |
| *Coronilla scorpioides* | Aerial parts | Ethyl acetate extracts | [135] |
| *Eremanthus incanus* | Aerial parts | Ethyl acetate extracts | [135] |
| *Pseudogynoxys cabrerae* | Aerial parts | Ethyl acetate extracts | [135] |
| *Leonotis nepetifolia* (L.) Br. | Roots and leaves | Hydroethanolic extracts | [136] |
| *Tanacetum parthenium* (L.) | Aerial parts | Dichloromethane fraction of hydroalcoholic extract | [131] |
| *Selaginella sellowii* | N/A | Hydroethanolic extract | [137] |
| *Selaginella sellowii* | N/A | Amentoflavone and robustaflavone | [137] |
| *Piper hispidum* | Leaves | 2'-Hydroxy-3′,4′,6′-trimethoxychalcone | [138] |

**Table 8.**
*Antileishmanial activity of extracts and isolated compounds against* L. amazonensis *amastigotes.*

and isolated compounds, phaeophytin A, 132(S)-hydroxyphaeophytin A, 132(S)-hydroxy-(173)-ethoxyphaeophorbide A and 7,4 -di-O-methylisoescutellarein, from aerial parts of *Sidastrum paniculatum* (L.) [141] exhibited antileishmanial activity against *L. braziliensis* promatsigotes (**Table 9**). Moreover, newly isolated compunds from *Delphinium gracile* [64] exhibited antileishmanial activity against *L. braziliensis* amastigotes (**Table 9**).

Isolated physalins from *Physalis minima* [42], methanol/water extract of leaves of *Otostegia limbata* (Benth.) [142], n-hexane extract of *Isodon rugosus* Wall. ex Benth [143], an isolated terpenoid, 3-O-Methyl-6-oxopristimerol, from root bark of *Maytenus chubutensis* [144], aqueous and methanolic extracts of *Capparis spinosa* L. [145], hydroalcoholic extract of leaves of *Avicennia marina* [146], methanol extract and ethyl acetate fraction of *Jurinea dolomiaea* [147], methanol/water extract of roots of *Malvastrum coromandelianum* (L.) Garcke [148], n-hexane extract of leaves of *Rhus Punjabensis* Stewart [149], and ethyl acetate extract of roots of *Dysphania ambrosioides* (L.) [150] exhibited antileishmanial activity against *L. tropica* promatsigotes (**Table 10**). In addition, neem seed oil of *Azadirachta indica* A. Juss [119] exhibited antileishmanial activity against *L. tropica* amastigotes (**Table 10**).

An isolated benzechromene from roots of *Bourreria pulchra* [151], an isolated compound, bokkosin, from roots of *Calliandra portoricensis* [152], essential oil and n-hexane extract of *Plagiochila porelloides* [153], isolated compounds, cycloartocarpesin and 1-O-linolenoyl-2-O-stearoyl-3-O-b-D-galactopyranosyl glycerol, from rhizomes of *Dorstenia contrajerva* [154], and an isolated compound, most active one: 2′,4′-dimethoxy-6′-hydroxychalcone, from aerial parts of *Polygonum salicifolium* [155] exhibited

| **L. braziliensis promatsigotes** | | | |
|---|---|---|---|
| **Plant** | **Plant part** | **Extract/isolated compound(s)** | **Reference** |
| *Ocotea lancifolia* | Stem bark | (−) Caaverine | [57] |
| *Delphinium gracile* | N/A | Newly identified | [64] |
| *Consolida oliveriana* | N/A | Newly identified | [64] |
| *Aconitum napellus* subsp. Lusitanicum | N/A | Newly identified | [64] |
| *Piper pseudoarboreum* | Leaves | 3-Chlorosintenpyridone and (S)-13-hydroxy-octadeca-(9Z,11E,15Z)-trienoic acid, Deacetylspathelin and casimiroin | [114] |
| *Annona mucosa* | Leaves | Liriodenine | [129] |
| *Trixis antimenorrhoea* | Aerial parts | Trixanolide | [130] |
| *Tarenaya spinosa* | Leaves | Ethanolic extract | [139] |
| *Pityrogramma calomelanos* (L.) | Leaves | Ethanol extract | [140] |
| *Sidastrum paniculatum* (L.) | Aerial parts | Phaeophytin A, 132(S)-hydroxyphaeophytin A, 132(S)- hydroxy-(173)-ethoxyphaeophorbide A, and 7,4 -di-O-methylisoescutellarein | [141] |
| *L. braziliensis* amastigotes | | | |
| *Delphinium gracile* | N/A | Newly identified | [64] |

**Table 9.**
*Antileishmanial activity of extracts and isolated compounds against* L. braziliensis *promatsigotes and amastigotes.*

| **L. tropica promatsigotes** | | | |
|---|---|---|---|
| **Plant** | **Plant part** | **Extract/isolated compound(s)** | **Reference** |
| *Physalis minima* | N/A | Isolated compounds | [42] |
| *Otostegia limbata* (Benth.) | Leaves | Methanol/water extract | [143] |
| *Isodon rugosus* Wall. ex Benth | N/A | n-Hexane extract | [143] |
| *Maytenus chubutensis* | Root bark | 3-O-Methyl-6-oxopristimerol | [144] |
| *Capparis spinosa* L. | N/A | Aqueous and methanolic extracts | [145] |
| *Avicennia marina* | Leaves | Hydroalcoholic extract | [146] |
| *Jurinea dolomiaea* | N/A | Methanol extract and ethyl acetate fraction | [147] |
| *Malvastrum coromandelianum* (L.) Garcke | Roots | Methanol/water extract | [148] |
| *Rhus Punjabensis* Stewart | Leaves | n-Hexane extract | [149] |
| *Dysphania ambrosioides* (L.) | Roots | Ethyl acetate extract | [150] |
| | *L. tropica* amastigotes | | |
| *Azadirachta indica* A. Juss | | Neem seed oil | [119] |

**Table 10.**
*Antileishmanial activity of extracts and isolated compounds against* L. tropica *promatsigotes and amastigotes.*

| ***L. mexicana* promatsigotes** | | | |
|---|---|---|---|
| **Plant** | **Plant part** | **Extract/isolated compound(s)** | **Reference** |
| *Bourreria pulchra* | Roots | Isolated compound | [151] |
| *Calliandra portoricensis* | Roots | Bokkosin | [152] |
| *Plagiochila porelloides* | N/A | Essential oil and n-hexane extract | [153] |
| *Dorstenia contrajerva* | Rhizomes | Cycloartocarpesin and 1-O- linolenoyl-2-O-stearoyl-3-O-b-D-galactopyranosyl glycerol | [154] |
| *Polygonum salicifolium* | Aerial parts | 2′,4'-Dimethoxy-6′-hydroxychalcone | [155] |
| *Thouinia paucidentata* Radlk | Leaves | n-Hexane extract | [156] |
| | | *L. mexicana* amastigotes | |
| *Thouinia paucidentata* Radlk | Leaves | n-Hexane extract | [156] |

**Table 11.**
*Antileishmanial activity of extracts and isolated compounds against* L. mexicana *promatsigotes and amastigotes.*

antileishmanial activity against *L. mexicana* promatsigotes (**Table 11**). Moreover, n-hexane extract of of leaves of *Thouinia paucidentata* Radlk [156] exhibited antileishmanial activity against *L. mexicana* both promatsigotes and amastigotes (**Table 11**).

Dichloromethane extract of earial parts of *Mitracarpus frigidus* [126] exhibited antileishmanial activity against *L. chagasi* promatsigotes, while ethyl acetate/methanol extract of leaves of *Spondias mombin* [157] exhibited antileishmanial activity against both *L. chagasi* promatsigotes and amastigotes (**Table 12**).

A hydromethanolic extract of leaves of *Clematis simensis fresen* and latex of *Euphorbia abyssinica* [72] exhibited antileishmanial activity against *L. aethiopica* promatsigotes (**Table 13**).

| ***L. chagasi* promatsigotes** | | | |
|---|---|---|---|
| **Plant** | **Plant part** | **Extract/isolated compound(s)** | **Reference** |
| *Mitracarpus frigidus* | Aerial parts | Dichloromethane extract | [126] |
| *Spondias mombin* | Leaves | Ethyl acetate/methanol extract | [157] |
| | | *L. chagasi* amastigotes | |
| *Spondias mombin* | Leaves | Ethyl acetate/methanol extract | [157] |

**Table 12.**
*Antileishmanial activity of extracts and isolated compounds against* L. chagasi *promatsigotes and amastigotes.*

| ***L. aethiopica* promatsigotes** | | | |
|---|---|---|---|
| **Plant** | **Plant part** | **Extract/isolated compound(s)** | **Reference** |
| *Euphorbia abyssinica* | Latex | Hydromethanolic extract | [72] |
| *Clematis simensis fresen* | Leaves | Hydromethanolic extract | [72] |

**Table 13.**
*Antileishmanial activity of extracts and isolated compounds against* L. aethiopica *promatsigotes.*

## 4. Conclusion

Recently, there has been a significant increase in scientific research aimed at developing effective treatments for leishmaniasis. Several investigations within the scientific community have highlighted the promising effects of phytochemicals as potential treatments for leishmaniasis. Based on these studies, extracts or essential oils obtained from *Spondias mombin, Physalis minima, Copaifera reticulate, Acacia nilotica, Ambrosia maratima, Vismia orientalis* Engl. *Capparis spinosa* L., *Zinowewia integerrima, Maytenus segoviarum, Quetzalia ilicina, Avicennia marina, Markhamia tomentosa* and medicinal *Piper* species, and phytochemicals form *Ancistrocladus tanzaniensis*, *Piper pseudoarboreum, Allanblackia monticola, Delphinium gracile, Consolida oliveriana* and *Aconitum napellus* subsp. Lusitanicum, *Ageratum conyzoides* L., *Crossostephium chinense*, and *Calea zacatechichi* are promising against different *Leishmania* species and should be further investigated.

Although several studies have shown promising results, most of the research has been limited to *in vitro* experiments utilizing promastigotes and amastigotes. Nonetheless, it is indispensable to conduct *in vivo* studies and standardize the techniques employed to evaluate compounds capable of surmounting the cellular barriers of host cells and withstanding the immune defenses of the body.

## Acknowledgements

The author acknowledges the use of Paperpal for grammar check and language polishing of the manuscript.

## Author details

Nergiz Gürbüz Çolak
Department of Molecular Biology and Genetics, İzmir Institute of Technology, İzmir, Türkiye

*Address all correspondence to: nergizgurbuz@iyte.edu.tr

**IntechOpen**

© 2024 The Author(s). Licensee IntechOpen. This chapter is distributed under the terms of the Creative Commons Attribution License (http://creativecommons.org/licenses/by/4.0), which permits unrestricted use, distribution, and reproduction in any medium, provided the original work is properly cited.

## References

[1] Di Bari C, Venkateswaran N, Fastl C, Gabriël S, Grace D, Havelaar AH, et al. The global burden of neglected zoonotic diseases: Current state of evidence. One Health. 2023;**17**:100595

[2] Welburn SC, Beange I, Ducrotoy MJ, Okello AL. The neglected zoonoses–the case for integrated control and advocacy. Clinical Microbiology and Infection. 2015;**21**(5):433-443

[3] World Health Organization. Ending the Neglect to Attain the Sustainable Development Goals: A Road Map for Neglected Tropical Diseases 2021-2030. Geneva, Switzerland: World Health Organization; 2021

[4] Alvar J, Vélez ID, Bern C, Herrero M, Desjeux P, Cano J, et al. Leishmaniasis worldwide and global estimates of its incidence. PLoS One. 2012;**7**(5):e35671

[5] Singh N, Mishra BB, Bajpai S, Singh RK, Tiwari VK. Natural product based leads to fight against leishmaniasis. Bioorganic & Medicinal Chemistry. 2014;**22**(1):18-45

[6] Stuart K, Brun R, Croft S, Fairlamb A, Gürtler RE, McKerrow J, et al. Kinetoplastids: Related protozoan pathogens, different diseases. The Journal of Clinical Investigation. 2008;**118**(4):1301-1310

[7] Maia C. Sand fly-borne diseases in Europe: Epidemiological overview and potential triggers for their emergence and re-emergence. Journal of Comparative Pathology. 2024;**209**:6-12

[8] Frézard F, Aguiar MMG, Ferreira LAM, Ramos GS, Santos TT, Borges GSM, et al. Liposomal amphotericin B for treatment of leishmaniasis: From the identification of critical physicochemical attributes to the design of effective topical and oral formulations. Pharmaceutics. 2022;**15**(1):99

[9] Mann S, Frasca K, Scherrer S, Henao-Martínez AF, Newman S, Ramanan P, et al. A review of Leishmaniasis: Current knowledge and future directions. Current Tropical Medicine Reports. 2021;**8**(2):121-132

[10] Pan American Health Organization (PAHO). Leishmaniasis; 2024. Available from: https://www.paho.org/en/topics/leishmaniasis [Accessed: July 2024]

[11] World Health Organization (WHO). Leishmaniasis; 2024. Available from: https://www.who.int/health-topics/leishmaniasis#tab=tab_1 [Accessed: July 2024]

[12] García-Estrada C, Pérez-Pertejo Y, Domínguez-Asenjo B, Holanda VN, Murugesan S, Martínez-Valladares M, et al. Further investigations of nitroheterocyclic compounds as potential antikinetoplastid drug candidates. Biomolecules. 2023;**13**(4):637

[13] Pinheiro AC, de Souza MVN. Current leishmaniasis drug discovery. RSC Medicinal Chemistry. 2022;**13**(9):1029-1043

[14] Elawad MA, Elkhalifa MEM, Hamdoon AAE, Salim LHM, Ahmad Z, Ayaz M. Natural products derived steroids as potential anti-leishmanial agents; disease prevalence, underlying mechanisms and future perspectives. Steroids. 2023;**193**:109196

[15] De Vries HJ, Reedijk SH, Schallig HD. Cutaneous leishmaniasis: Recent developments in diagnosis and

management. American Journal of Clinical Dermatology. 2015;**16**(2):99-109

[16] Lazar LTY, Abass KS. Morphology, life cycle, pathogenesis and virulence factors of genus Leishmania: A review. Plant Archives. 2020;**20**(2):4057-4060

[17] Cecílio P, Cordeiro-da-Silva A, Oliveira F. Sand flies: Basic information on the vectors of leishmaniasis and their interactions with *Leishmania* parasites. Communications Biology. 2022;**5**(1):305

[18] Bharadava K, Upadhyay TK, Kaushal RS, Ahmad I, Alraey Y, Siddiqui S, et al. Genomic insight of leishmania parasite: In-depth review of drug resistance mechanisms and genetic mutations. ACS Omega. 2024;**9**(11):12500-12514

[19] Alvar J, Aparicio P, Aseffa A, Den Boer M, Cañavate C, Dedet JP, et al. The relationship between leishmaniasis and AIDS: The second 10 years. Clinical Microbiology Reviews. 2008;**21**(2):334-359

[20] Oryan A, Akbari M. Worldwide risk factors in leishmaniasis. Asian Pacific Journal of Tropical Medicine. 2016;**9**(10):925-932

[21] Kevric I, Cappel MA, Keeling JH. New world and old world Leishmania infections: A practical review. Dermatologic Clinics. 2015;**33**(3):579-593

[22] Wortmann G, Aronson N, Miller RS, Jackson J, Oster C. A randomized, double-blind study of the efficacy of a 10- or 20-day course of sodium stibogluconate for treatment of cutaneous leishmaniasis in United States military personnel. Clinical Infectious Diseases. 2002;**35**(3):261-267

[23] Haldar AK, Sen P, Roy S. Use of antimony in the treatment of leishmaniasis: Current status and future directions. Molecular Biology International. 2011;**2011**:571242

[24] Singh R, Kashif M, Srivastava P, Manna PP. Recent advances in chemotherapeutics for Leishmaniasis: Importance of the cellular biochemistry of the parasite and its molecular interaction with the host. Pathogens. 2023;**12**(5):706

[25] Marques SA, Merlotto MR, Ramos PM, Marques MEA. American tegumentary leishmaniasis: Severe side effects of pentavalent antimonial in a patient with chronic renal failure. Anais Brasileiros de Dermatologia. 2019;**94**(3):355-357

[26] Hamill RJ. Amphotericin B formulations: A comparative review of efficacy and toxicity. Drugs. 2013;**73**(9):919-934

[27] De Menezes JP, Guedes CE, Petersen AL, Fraga DB, Veras PS. Advances in development of new treatment for leishmaniasis. BioMed Research International. 2015;**2015**:815023

[28] Laniado-Laborín R, Cabrales-Vargas MN. Amphotericin B: Side effects and toxicity. Revista Iberoamericana de Micología. 2009;**26**(4):223-227

[29] Dupont B. Overview of the lipid formulations of amphotericin B. The Journal of Antimicrobial Chemotherapy. 2002;**49**(Suppl. 1):31-36

[30] Romero GAS, Costa DL, Costa CHN, de Almeida RP, de Melo EV, et al. Efficacy and safety of available treatments for visceral leishmaniasis in Brazil: A multicenter, randomized, open label trial. PLoS Neglected Tropical Diseases. 2017;**11**(6):e0005706

[31] Faustino C, Pinheiro L. Lipid systems for the delivery of amphotericin B in

antifungal therapy. Pharmaceutics. 2020;**12**(1):29

[32] Dorlo TP, Balasegaram M, Beijnen JH, de Vries PJ. Miltefosine: A review of its pharmacology and therapeutic efficacy in the treatment of leishmaniasis. The Journal of Antimicrobial Chemotherapy. 2012;**67**(11):2576-2597

[33] Sundar S, Singh A, Rai M, Prajapati VK, Singh AK, Ostyn B, et al. Efficacy of miltefosine in the treatment of visceral leishmaniasis in India after a decade of use. Clinical Infectious Diseases. 2012;**55**(4):543-550

[34] Croft SL, Engel J. Miltefosine–discovery of the antileishmanial activity of phospholipid derivatives. Transactions of the Royal Society of Tropical Medicine and Hygiene. 2006;**100**(Suppl. 1):S4-S8

[35] Sundar S, Jha TK, Thakur CP, Sinha PK, Bhattacharya SK. Injectable paromomycin for visceral leishmaniasis in India. The New England Journal of Medicine. 2007;**356**(25):2571-2581

[36] Andersen EM, Cruz-Saldarriaga M, Llanos-Cuentas A, Luz-Cjuno M, Echevarria J, Miranda-Verastegui C, et al. Comparison of meglumine antimoniate and pentamidine for peruvian cutaneous leishmaniasis. The American Journal of Tropical Medicine and Hygiene. 2005;**72**(2):133-137

[37] Brindha J, Balamurali MM, Chanda K. An overview on the therapeutics of neglected infectious diseases-leishmaniasis and chagas diseases. Frontiers in Chemistry. 2021;**9**:622286

[38] Singh K, Garg G, Ali V. Current therapeutics, their problems and thiol metabolism as potential drug targets in leishmaniasis. Current Drug Metabolism. 2016;**17**(9):897-919

[39] Scholar E. Paromomycin. In: xPharm: The Comprehensive Pharmacology Reference. Amsterdam: Elsevier; 2008. pp. 1-4

[40] Sundar S, Chakravarty J. Paromomycin in the treatment of leishmaniasis. Expert Opinion on Investigational Drugs. 2008;**17**(5):787-794

[41] Bamorovat M, Sharifi I, Khosravi A, Aflatoonian MR, Agha Kuchak Afshari S, Salarkia E, et al. Global dilemma and needs assessment toward achieving sustainable development goals in controlling leishmaniasis. Journal of Epidemiology and Global Health. 2024;**14**(1):22-34

[42] Bano S, Bibi M, Farooq S, Zafar H, Shaikh M, Khoso BK, et al. Anti-leishmanial physalins-phytochemical investigation, in vitro evaluation against clinical and MIL-resistant *L. tropica* strains and in silico studies. PLoS One. 2022;**17**(11):e0274543

[43] Samuelsson G. Drugs of natural origin: A textbook of Pharmacognosy. In: 5th Swedish Pharmaceutical Press. Stockholm: ACS Publications; 2004

[44] Salim A, Chin YW, Kinghorn A. Drug discovery from plants. In: Ramawat K, Merillon J, editors. Bioactive Molecules and Medicinal Plants. Berlin, Heidelberg: Springer; 2008. pp. 1-24

[45] Al-Huqail AA, Bekhit AA, Mostafa NM, Ullah H, Ayaz M. Antimalarial and antileishmanial flavonoids from *Calendula officinalis* flowers. Agronomy (Basel, Switzerland). 2023;**13**(11):2765

[46] Dias DA, Urban S, Roessner U. A historical overview of natural products in drug discovery. Metabolites. 2012;**2**(2):303-336

[47] Iwu MM, Jackson JE, Schuster BG. Medicinal plants in the fight against leishmaniasis. Parasitology Today. 1994;**10**(2):65-68

[48] Sebisubi FM, Tan GT. Phytochemistry and Pharmacognosy: Natural Products with Antimalarial Activity, Encyclopedia of Life Support Systems (EOLSS). Paris, France: UNESCO-EOLSS; 2016

[49] Ayaz M, Ullah F, Sadiq A, Kim MO, Ali T. Editorial: Natural products-based drugs: Potential therapeutics against alzheimer's disease and other neurological disorders. Frontiers in Pharmacology. 2019;**10**:1417

[50] Ayaz M, Ahmad I, Sadiq A, Ullah F, Ovais M, Khalil AT, et al. *Persicaria hydropiper* (L.) Delarbre: A review on traditional uses, bioactive chemical constituents and pharmacological and toxicological activities. Journal of Ethnopharmacology. 2020;**251**:112516

[51] Bernardini S, Tiezzi A, Laghezza Masci V, Ovidi E. Natural products for human health: An historical overview of the drug discovery approaches. Natural Product Research. 2018;**32**(16):1926-1950

[52] Siddiqui AA, Iram F, Siddiqui S, Sahu K. Role of natural products in drug discovery process. International Journal of Drug Development and Research. 2014;**6**(2):172-204

[53] Abdel-Mageed WM, Backheet EY, Khalifa AA, Ibraheim ZZ, Ross SA. Antiparasitic antioxidant phenylpropanoids and iridoid glycosides from *Tecoma mollis*. Fitoterapia. 2012;**83**(3):500-507

[54] Cheikh-Ali Z, Okpekon T, Roblot F, Bories C, Cardao M, Jullian JC, et al. Labdane diterpenoids from *Aframomum sceptrum*: NMR study and antiparasitic activities. Phytochemistry Letters. 2011;**4**(3):240-244

[55] Das A, Das MC, Das N, Bhattacharjee S. Evaluation of the antileishmanial potency, toxicity and phytochemical constituents of methanol bark extract of *Sterculia villosa*. Pharmaceutical Biology. 2017;**55**(1):998-1009

[56] Eltayeb A, Ibrahim K. Potential antileishmanial effect of three medicinal plants. Indian Journal of Pharmaceutical Sciences. 2012;**74**(2):171-174

[57] Fournet A, Ferreira ME, Rojas de Arias A, Guy I, Guinaudeau H, Heinzen H. Phytochemical and antiprotozoal activity of *Ocotea lancifolia*. Fitoterapia. 2007;**78**(5):382-384

[58] Ghosh J, Chakraborty S, Dey S, Mukherjee D, Sarkar B, Mallick S, et al. Potential anti-leishmanial activity of a semi-purified fraction isolated from the leaves of *Parthenium hysterophorus*. Acta Parasitologica. 2021;**66**(4):1480-1489

[59] Guetchueng ST, Djouonzo PT, Lame Y, Kopa Kowa T, Dotse E, et al. Antileishmanial anthraquinones from the rhyzomes of *Rumex abyssinicus* Jacq (Polygonaceae). Natural Product Research. 2023;**37**(17):2935-2939

[60] Hata Y, De Mieri M, Ebrahimi SN, Mokoka T, Fouche G, Kaiser M, et al. Identification of two new phenathrenones and a saponin as antiprotozoal constituents of *Drypetes gerrardii*. Phytochemistry Letters. 2014;**28**(10):cxxxiii-cxl

[61] Keshav P, Goyal DK, Kaur S. Promastigotes of Leishmania donovani exhibited sensitivity towards the high altitudinal plant *Cicer microphyllum*. Current Research in Parasitology & Vector-borne Diseases. 2021;**1**:100040

[62] Ma'mag LK, Zintchem AAA, Théodora KK, Atchadé AT, Lauve TY, Frédérich M, et al. Antiplasmodial and antileishmanial inhibitory activity of triterpenes and steroidal alkaloid from the leaves of *Funtumia elastica* (Preuss) Stapf (Apocynaceae). Fitoterapia. 2021;**151**:104869

[63] Kom Ma'Mag L, AAA Z, Tchinda A, Kopa Kowa T, Frédérich M, Ngono Bikobo DS, et al. Funtulaticamide, a phytosphingosine-type ceramide from *Funtumia elastica* Preuss Stapf. (Apocynaceae) trunk bark with potential antileishmanial activity. Biochemical Systematics and Ecology. 2022;**106**:104569

[64] Marín C, Díaz JG, Irure Maiques D, Ramírez-Macías I, Rosales MJ, Guitierrez-Sánchez R, et al. Antitrypanosomatid activity of flavonoid glycosides isolated from *Delphinium gracile*, *D. Staphisagria*, *Consolida oliveriana* and from *Aconitum napellus* subsp. Lusitanicum. Phytochemistry Letters. 2017;**19**(19):196-209

[65] Moawad A, Hetta M, Zjawiony JK, Ferreira D, Hifnawy M. Two new dihydroamentoflavone glycosides from *Cycas revoluta*. Natural Product Research. 2014;**28**(1):41-47

[66] Mohamed NM, Makboul MA, Farag SF, Jain S, Jacob MR, Tekwani BL, et al. Triterpenes from the roots of *Lantana montevidensis* with antiprotozoal activity. Phytochemistry Letters. 2015;**15**:30-36

[67] Mostafa AE, El-Hela AA, Mohammad AEI, Jacob M, Cutler SJ, Ross SA. New secondary metabolites from *Dodonaea viscosa*. Phytochemistry Letters. 2014;**8**:10-15

[68] Núñez MJ, Martínez ML, Castillo UG, Flores KC, Menjívar J, et al. Salvadoran *Celastraceae* species as a source of antikinetoplastid quinonemethide triterpenoids. Plants (Basel). 2024;**13**(3):360

[69] Pyne N, Paul S. Screening of medicinal plants unraveled the leishmanicidal credibility of *Garcinia cowa*; highlighting Norcowanin, a novel anti-leishmanial phytochemical through in-silico study. Journal of Parasitic Diseases. 2022;**46**(1):202-214

[70] Sawadogo WR, Le Douaron G, Maciuk A, Bories C, Loiseau PM, Figadère B, et al. In vitro antileishmanial and antitrypanosomal activities of five medicinal plants from Burkina Faso. Parasitology Research. 2012;**110**(5):1779-1783

[71] Utage B, Patole M, Nagvenkar P, Gacche R. Prosopis juliflora (Sw.) DC.Induces apoptotic-like programmed cell death in Leishmania donovani via over production of oxidative stress, mitochondrial dysfunction and ATP depletion. Journal of Traditional and Complementary Medicine. 2023;**13**(6):611-622

[72] Worku KM, Araya D, Tesfa H, Birru EM, Hailu A, Aemero M. In vitro antileishmanial activities of hydro-methanolic crude extracts and solvent fractions of *Clematis simensis* fresen leaf, and *Euphorbia abyssinica* latex. Medicine (Baltimore). 2024;**103**(18):e38039

[73] Wu H, Fronczek FR, Burandt CL Jr, Zjawiony JK. Antileishmanial germacranolides from *Calea zacatechichi*. Planta Medica. 2011;**77**(7):749-753

[74] Afolayan M, Srivedavyasasri R, Asekun OT, Familoni OB, Orishadipe A, Zulfiqar F, et al. Phytochemical study of *Piliostigma thonningii*, a medicinal plant grown in Nigeria. Medicinal Chemistry Research. 2018;**27**(10):2325-2330

[75] Akaberi M, Najaran ZT, Azizi N, Emami SA. Metabolite profiling and antiprotozoal activity of three endemic Iranian *Helichrysum* species. Industrial Crops and Products. 2021;**174**:114196

[76] Bringmann G, Wohlfarth M, Rischer H, Schlauer J, Brun R. Extract screening by HPLC coupled to MS-MS, NMR, and CD: A dimeric and three monomeric naphthylisoquinoline alkaloids from *Ancistrocladus griffithii*. Phytochemistry. 2002;**61**(2):195-204

[77] Bringmann G, Dreyer M, Faber JH, Dalsgaard PW, Staerk D, Jaroszewski JW, et al. Ancistrotanzanine A, the first 5,3′-coupled naphthylisoquinoline alkaloid, and two further, 5,8′-linked related compounds from the newly described species *Ancistrocladus tanzaniensis*. Journal of Natural Products. 2003;**66**(9):1159-1165

[78] Farimani MM, Khodaei B, Moradi H, Aliabadi A, Ebrahimi SN, De Mieri M, et al. Phytochemical study of salvia leriifolia roots: Rearranged abietane diterpenoids with antiprotozoal activity. Journal of Natural Products. 2018;**81**(6):1384-1390

[79] Ganapaty S, Pannakal ST, Srilakshmi GVK, Lakshmi P, Waterman PG, Brun R. Pumilanol, an antiprotozoal isoflavanol from *Tephrosia pumila*. Phytochemistry Letters. 2008;**1**(4):175-178

[80] Hay AE, Ioset JR, Ahua KM, Diallo D, Brun R, Hostettmann K. Limonoid orthoacetates and antiprotozoal compounds from the roots of *Pseudocedrela kotschyi*. Journal of Natural Products. 2007;**70**(1):9-13

[81] Kirmizibekmez H, Atay I, Kaiser M, Brun R, Cartagena MM, Carballeira NM, et al. Antiprotozoal activity of *Melampyrum arvense* and its metabolites. Phytotherapy Research. 2011;**25**(1):142-146

[82] Kunert O, Swamy RC, Kaiser M, Presser A, Buzzi S, Appa Rao AVN, et al. Antiplasmodial and leishmanicidal activity of biflavonoids from Indian *Selaginella bryopteris*. Phytochemistry Letters. 2008;**1**(4):171-174

[83] Lagnika L, Attioua B, Weniger B, Kaiser M, Sanni A, Vonthron-Senecheau C. Phytochemical study and antiprotozoal activity of compounds isolated from *Thalia geniculata*. International Journal of Pharmacognosy. 2008;**46**(3):162-165

[84] Lagnika L, Weniger B, Attioua B, Jensen O, Anthaume C, Sanni A, et al. Trypanocidal activity of diarylheptanoids from *Schrankia leptocarpa* DC. South African Journal of Botany. 2012;**83**:92-97

[85] Lenta BN, Vonthron-Sénécheau C, Weniger B, Devkota KP, Ngoupayo J, Kaiser M, et al. Leishmanicidal and cholinesterase inhibiting activities of phenolic compounds from *Allanblackia monticola* and *Symphonia globulifera*. Molecules. 2007;**12**(8):1548-1557

[86] López D, Cherigo L, Spadafora C, Loza-Mejía MA, Martínez-Luis S. Phytochemical composition, antiparasitic and α-glucosidase inhibition activities from *Pelliciera rhizophorae*. Chemistry Central Journal. 2015;**9**(1):53

[87] Mbwambo ZH, Apers S, Moshi MJ, Kapingu MC, Van Miert S, Claeys M, et al. Anthranoid compounds with antiprotozoal activity from *Vismia orientalis*. Planta Medica. 2004;**70**(8):706-710

[88] Nour AM, Khalid SA, Kaiser M, Brun R, Abdallah WE, Schmidt TJ. The antiprotozoal activity of sixteen asteraceae species native to Sudan

and bioactivity-guided isolation of xanthanolides from *xanthium brasilicum*. Planta Medica. 2009;**75**(12):1363-1368

[89] Nour AM, Khalid SA, Kaiser M, Brun R, Abdalla WE, Schmidt TJ. The antiprotozoal activity of methylated flavonoids from *Ageratum conyzoides* L. Journal of Ethnopharmacology. 2010;**129**(1):127-130

[90] Schmidt TJ, Rzeppa S, Kaiser M, Brun R. *Larrea tridentata*—Absolute configuration of its epoxylignans and investigations on its antiprotozoal activity. Phytochemistry Letters. 2012;**5**(3):632-638

[91] Tantangmo F, Lenta BN, Boyom FF, Ngouela S, Kaiser M, Tsamo E, et al. Antiprotozoal activities of some constituents of *Markhamia tomentosa* (Bignoniaceae). Annals of Tropical Medicine and Parasitology. 2010;**104**(5):391-398

[92] Vásquez-Ocmín PG, Gadea A, Cojean S, Marti G, Pomel S, Van Baelen AC, et al. Metabolomic approach of the antiprotozoal activity of medicinal *piper* species used in Peruvian Amazon. Journal of Ethnopharmacology. 2021;**264**:113262

[93] Zadali R, Nejad Ebrahimi S, Tofighi Z, Es-Haghi A, Hamburger M, Kaiser M, et al. Antiprotozoal activity of diterpenoids isolated from *Zhumeria majdae*- absolute configuration by circular dichroism. Daru. 2020;**28**(2):455-462

[94] Al Nasr I. In vitro anti-leishmanial assessment of some medicinal plants collected from Al Qassim, Saudi Arabia. Acta Parasitologica. 2020;**65**(3):696-703

[95] Sousa E, Chaves M, Santos I, Sousa V, Júnior G, Rosa A, et al. composição química e atividade antileishmania de *Tocoyena hispidula*. Quimica Nova. 2018;**42**(2):1481-1484

[96] Devkota KP, Lenta BN, Wansi JD, Choudhary MI, Kisangau DP, Naz Q, et al. Bioactive 5alpha-pregnane-type steroidal alkaloids from *Sarcococca hookeriana*. Journal of Natural Products. 2008;**71**(8):1481-1484

[97] Fattahian M, Ghanadian M, Zolfaghari B, Abdeyazdan S, Saberi S, Zulfiqar F, et al. Phytochemical study of *Seriphidium khorassanicum* (syn. Artemisia khorassanica) aerial parts: Sesquiterpene lactones with anti-protozoal activity. Natural Product Research. 2024;**38**(1):16-27

[98] Kefi S, Essid R, Mkadmini K, Kefi A, Mahjoub Haddada F, Tabbene O, et al. Phytochemical investigation and biological activities of *Echium arenarium* (Guss) extracts. Microbial Pathogenesis. 2018;**118**:202-210

[99] Khan TA, Al Nasr IS, Mujawah AH, Koko WS. Assessment of *Euphorbia retusa* and *Pulicaria undulata* activity against *Leishmania major* and *toxoplasma gondii*. Tropical Biomedicine. 2021;**38**(1):135-141

[100] Mahmoud AH, Mahmoud BK, Samy MN, Fouad MA, Kamel MS, Matsunami K. Cytotoxic and antileishmanial triterpenes of *Tabebuia aurea* (Silva Manso) leaves. Natural Product Research. 2022;**36**(23):6181-6185

[101] Morel S, Helesbeux JJ, Séraphin D, Derbré S, Gatto J, Aumond MC, et al. Anti-AGEs and antiparasitic activity of an original prenylated isoflavonoid and flavanones isolated from Derris ferruginea. Phytochemistry Letters. 2013;**6**(3):498-503

[102] Rezaee F, Zolfaghari B, Dinani MS. Isolation of dioscin-related steroidal saponin from the bulbs of *Allium*

*paradoxum* L. with leishmanicidal activity. Research in Pharmaceutical Sciences. 2018;**13**(5):469-475

[103] Sakunpak A, Matsunami K, Otsuka H, Panichayupakaranant P. Isolation of new monoterpene coumarins from *Micromelum minutum* leaves and their cytotoxic activity against *Leishmania major* and cancer cells. Food Chemistry. 2013;**139**(1-4):458-463

[104] Sakunpak A, Panichay-upakaranant P, Otsuka H, Matsunami K. Isolation of chamuangone, a cytotoxic compound against *Leishmania major* and cancer cells from *Garcinia cowa* leaves and its HPLC quantitative determination method. Journal of Cancer Research Updates. 2017;**6**(2):38-45

[105] Sharifi-Rad J, Roointan A, Setzer WN, Sharifi-Rad M, Iriti M, Salehi B. Susceptibility of *Leishmania major* to *Veronica persica* Poir. Extracts - In vitro and in vivo assays. Cellular and Molecular Biology (Noisy-le-Grand, France). 2018;**64**(8):44-49

[106] Sharifi F, Sharififar F, Pournamdari M, Ansari M, Tavakoli Oliaee R, Bamorovat M, et al. Leishmanicidal potentials of *Gossypium hirsutum* extract and its fractions on *Leishmania major* in a murine model: Parasite burden, gene expression, and histopathological profile. Journal of Medical Microbiology. 2021;**70**(6)

[107] Tahir AE, Ibrahim AM, Satti GM, Theander TG, Kharazmi A, Khalid SA. The potential antileishmanial activity of some Sudanese medicinal plants. Phytotherapy Research. 1998;**12**(8):576-579

[108] Uddin G, Rauf A, Arfan M, Waliullah, Khan I, Ali M, et al. Pistagremic acid a new leishmanicidal triterpene isolated from *Pistacia integerrima* Stewart. Journal of Enzyme Inhibition and Medicinal Chemistry. 2012;**27**(5):646-648

[109] Wang Z, Chitama BA, Suganuma K, Yamano Y, Sugimoto S, Kawakami S, et al. Two new cytotoxic sesquiterpene-amino acid conjugates and a coumarin-glucoside from *Crossostephium chinense*. Molecules. 2023;**28**(12):4696

[110] Wang Z, Yamano Y, Kawakami S, Al-Hamoud GA, Sugimoto S, Otsuka H, et al. New ψ-santonin derivatives from *Crossostephium chinense* and their anti-proliferative activities against *Leishmania major* and human cancer cells A549. Molecules. 2023;**28**(24):8108

[111] Albalawi AE. Antileishmanial activity of *Ziziphus spina*-christi leaves extract and its possible cellular mechanisms. Microorganisms. 2021;**9**(10):2113

[112] Aqeele G, Shayan P, Ebrahimzade Abkooh E, Mohebali M. Evaluation of curcumin and CM11 peptide alone and in combination against amastigote form of Iranian strain of L. major (MRHO/IR75/ER) in vitro. Experimental Parasitology. 2021;**229**:108151

[113] Cunha LF, Mendonça Filho CV, Campos FF, Grael CFF, Oliveira FD, Barroso PR, et al. Phytochemical screening and biological assays of ethanolic leaf extract of *Senna rugosa*. Rodriguésia. 2020;**71**

[114] Flores N, Ticona JC, Bilbao-Ramos P, Dea-Ayuela MA, Ruiz Macedo JC, et al. An unprecedented chlorine-containing piperamide from *piper pseudoarboreum* as potential leishmanicidal agent. Fitoterapia. 2019;**134**:340-345

[115] Mazoir N, Benharref A, Bailén M, Reina M, González-Coloma A, Martínez-Díaz RA. Antileishmanial and

antitrypanosomal activity of triterpene derivatives from latex of two *euphorbia species*. Zeitschrift Für Naturforschung. C, A Journal of Biosciences. 2011;**66**(7-8):360-366

[116] Ramadani AP, Paloque L, Belda H, Tamhid HA, Masriani M, Jumina J, et al. Antiprotozoal properties of Indonesian medicinal plant extracts. Journal of Herbal Medicine. 2017;**11**:46-52

[117] Al-Massarani SM, El Gamal AA, Al-Musayeib NM, Mothana RA, Basudan OA, Al-Rehaily AJ, et al. Phytochemical, antimicrobial and antiprotozoal evaluation of *Garcinia mangostana* pericarp and α-mangostin, its major xanthone derivative. Molecules. 2013;**18**(9):10599-10608

[118] Bosquiroli LSS, Demarque DP, Rizk YS, Cunha MC, Marques MCS, Matos MDFC, et al. In vitro anti-Leishmania infantum activity of essential oil from *Piper angustifolium*. Revista Brasileira de Farmacognosia. 2015;**25**(2):124-128

[119] Cesa S, Sisto F, Zengin G, Scaccabarozzi D, Kokolakis AK, Scaltrito MM, et al. Phytochemical analyses and pharmacological screening of neem oil. South African Journal of Botany. 2018;**120**:331-337

[120] Al Musayeib NM, Mothana RA, Mohamed GA, Ibrahim SR, Maes L. Hypoestenonols A and B, new fusicoccane diterpenes from *Hypoestes forskalei*. Phytochemistry Letters. 2014;**10**:23-27

[121] Andrade E, Soares M, Silva D, Neta L, Júnior A, Guimarães E, et al. Phytochemical analysis, multi-element composition and biological activities of extracts and lupenone from *Albizia inundata* (Mart.) Barneby & J.W.Grimes. Journal of the Brazilian Chemical Society. 2020;**31**(9):1825-1837

[122] Comandolli-Wyrepkowski CD, Santos PAD, Grafov A, Grafova I, Franco AMR, Silva AFD, et al. Antileishmanial activity of extracts from *Libidibia ferrea*: Development of in vitro and in vivo tests. Acta Amazonica. 2017;**47**(4):331-340

[123] Santos AO, Ueda-Nakamura T, Dias Filho BP, Veiga Junior VF, Pinto AC, Nakamura CV. Effect of Brazilian copaiba oils on *Leishmania amazonensis*. Journal of Ethnopharmacology. 2008;**120**(2):204-208

[124] Dos Santos AO, Izumi E, Ueda-NakamuraT, Dias-FilhoBP, Veiga-JúniorVF, Nakamura CV. Antileishmanial activity of diterpene acids in copaiba oil. Memórias do Instituto Oswaldo Cruz. 2013;**108**(1):59-64

[125] Ellafi A, Farhat R, Snoussi M, Noumi E, Anouar EH, Ben Ali R, et al. Phytochemical profiling, antimicrobial, antibiofilm, insecticidal, and anti-leishmanial properties of aqueous extract from *Juglans regia* L. root bark: In vitro and in silico approaches. International Journal of Food Properties. 2023;**26**(1):1079-1097

[126] Fabri RL, Nogueira MS, Braga FG, Coimbra ES, Scio E. *Mitracarpus frigidus* aerial parts exhibited potent antimicrobial, antileishmanial, and antioxidant effects. Bioresource Technology. 2009;**100**(1):428-433

[127] Fernandes PAS, Silva JCPD, Lima Sales D, Ribeiro PRV, Sousa de Brito E, et al. Chemical constituents and biological activities of croton heliotropiifolius Kunth. Antibiotics (Basel). 2021;**10**(9):1074

[128] Lucilia K, Miranda N, Faria EO, Oliveira CMAD, Silva CCD, Iglesias AH, et al. Antiprotozoal alkaloids from *Psychotria prunifolia* (Kunth) Steyerm.

Journal of the Brazilian Chemical Society. 2012;**23**(2):355-360

[129] Lima JPSD, Santos AMG, Barison A, Pinheiro MLB, Costa EV, Pereira JLDS, et al. In vitro atileishmanial and cytotoxic activities of *Annona mucosa* (Annonaceae). Revista Virtual de Química. 2012;**4**(6):692-702

[130] Maldonado EM, Salamanca E, Giménez A, Saavedra G, Sterner O. Antileishmanial metabolites from *Trixis antimenorrhoea*. Phytochemistry Letters. 2014;**10**:281-286

[131] Rabito MF, Britta EA, Pelegrini BL, Scariot DB, Almeida MB, Nixdorf SL, et al. In vitro and in vivo antileishmania activity of sesquiterpene lactone-rich dichloromethane fraction obtained from *Tanacetum parthenium* (L.) Schultz-Bip. Experimental Parasitology. 2014;**143**:18-23

[132] Taleb-Contini SH, De Oliveira DCR, Salvador MJ, Balanco JMF, Albuquerque S. Antiprotozoal effect of crude extracts and flavonoids isolated from *Chromolaena hirsuta* (asteraceae). Phytotherapy Research. 2004;**18**(3):250-254

[133] Azebaze AG, Ouahouo BM, Vardamides JC, Valentin A, Kuete V, Acebey L, et al. Antimicrobial and antileishmanial xanthones from the stem bark of *Allanblackia gabonensis* (Guttiferae). Natural Product Research. 2008;**22**(4):333-341. Retraction in: Nat Prod Res. 2010 May;24(8):781

[134] Cabanillas BJ, Le Lamer AC, Olagnier D, Castillo D, Arevalo J, Valadeau C, et al. Leishmanicidal compounds and potent PPARγ activators from *Renealmia thyrsoidea* (Ruiz & Pav.) Poepp. & Endl. Journal of Ethnopharmacology. 2014;**157**:149-155

[135] Da Silva ACN, Do Nascimento RMC, Rodrigues DCDN, Ferreira PMP, Pessoa C, Lima DJB, et al. In vitro activity evaluation of seven Brazilian *Asteraceae* against cancer cells and Leishmania amazonensis. South African Journal of Botany. 2018;**121**:267-273

[136] De Oliveira DP, de Almeida L, Marques MJ, de Carvalho RR, Dias ALT, da Silva GA, et al. Exploring the bioactivity potential of *Leonotis nepetifolia*: Phytochemical composition, antimicrobial and antileishmanial activities of extracts from different anatomical parts. Natural Product Research. 2021;**35**(18):3120-3125

[137] Rizk YS, Fischer A, Cunha Mde C, Rodrigues PO, Marques MC, Matos Mde F, et al. In vitro activity of the hydroethanolic extract and biflavonoids isolated from *Selaginella sellowii* on *Leishmania* (Leishmania) amazonensis. Memórias do Instituto Oswaldo Cruz. 2014;**109**(8):1050-1056

[138] Ruiz C, Haddad M, Alban J, Bourdy G, Reategui R, Castillo D, et al. Activity-guided isolation of antileishmanial compounds from *Piper hispidum*. Phytochemistry Letters. 2011;**4**(3):363-366

[139] Bezerra JWA, Coronel C, Gomez MCV, Rolón M, Nunez CV, Da Silva DR, et al. Evaluation of antiparasitary, cytotoxic and antioxidant activity and chemical analysis of *Tarenaya spinosa* (Jacq.) Raf. (Cleomaceae). South African Journal of Botany. 2019;**124**:546-555

[140] Souza TM, Morais-Braga MF, Saraiva AÁ, Rolón M, Vega C, de Arias AR, et al. Evaluation of the anti-Leishmania activity of ethanol extract and fractions of the leaves from *Pityrogramma calomelanos* (L.) link. Natural Product Research. 2013;**27**(11):992-996

[141] Teles YCF, Chaves OS, Agra MDF, Batista LM, Queiroz ACD, Araújo MVD,

et al. Chemical constituents from *Sidastrum paniculatum* and evaluation of their leishmanicidal activity. Revista Brasileira de Farmacognosia. 2015;**25**(4):363-368

[142] Fatima H, Kainat A, Akbar F, Shinwari ZK, Naz I. Polarity guided extraction, HPLC based phytochemical quantification, and multimode biological evaluation of *Otostegia limbata* (Benth.) Boiss. Arabian Journal of Chemistry. 2022;15(2):103583

[143] Hameed A, Zafar M, Ahmad M, Sultana S, Akhter MS, Zaman W, et al. Micromorphology, phytochemical and pharmacological evaluation of *Isodon rugosus* (wall. Ex Benth.) Codd. JAPS. 2022;**32**(3)

[144] Kennedy ML, Llanos GG, Castanys S, Gamarro F, Bazzocchi IL, Jiménez IA. Terpenoids from Maytenus species and assessment of their reversal activity against a multidrug-resistant *Leishmania tropica* line. Chemistry & Biodiversity. 2011;**8**(12):2291-2298

[145] Nazer MR, Jahanbakhsh S, Ebrahimi K, Niazi M, Sepahvand M, Khatami M, et al. Cytotoxic and antileishmanial effects of various extracts of *Capparis spinosa* L. Turkish Journal of Pharmaceutical Sciences. 2021;**18**(2):146-150

[146] Rashno Z, Sharifi I, Oliaee RT, Tajbakhsh S, Sharififar F, Sharifi F, et al. Anti-leishmanial activity of *Avicennia marina* (*Avicenniaceae* family) leaves hydroalcoholic extract and its possible cellular mechanisms. Parasite Epidemiology and Control. 2022;**17**:e00239

[147] Shah NA, Khan MR, Nadhman A. Antileishmanial, toxicity, and phytochemical evaluation of medicinal plants collected from Pakistan. BioMed Research International. 2014;**2014**:384204

[148] Sheema S, Salman Zafar SZ, Nazif Ullah NU, AGUD GUD, Ishaq Khan IK. Phytochemical and In vitro biological evaluation of roots of Malvastrum coromandelianum (L.) Garcke. Journal of the Chemical Society of Pakistan. 2023;**45**(6):568

[149] Tabassum S, Ahmed M, Mirza B, Naeem M, Zia M, Shanwari ZK, et al. Appraisal of phytochemical and in vitro biological attributes of an unexplored folklore: *Rhus Punjabensis* Stewart. BMC Complementary and Alternative Medicine. 2017;**17**(1):146

[150] Zohra T, Ovais M, Khalil AT, Qasim M, Ayaz M, Shinwari ZK. Extraction optimization, total phenolic, flavonoid contents, HPLC-DAD analysis and diverse pharmacological evaluations of *Dysphania ambrosioides* (L.) Mosyakin & Clemants. Natural Product Research. 2019;**33**(1):136-142

[151] Erosa-Rejón GJ, Yam-Puc A, Chan-Bacab MJ, Giménez-Turbax A, Salamanca E, Peña-Rodríguez LM, et al. Benzochromenes from the roots of *Bourreria pulchra*. Phytochemistry Letters. 2009;**3**(1):9-12

[152] Nvau JB, Alenezi S, Ungogo MA, Alfayez IAM, Natto MJ, et al. Antiparasitic and cytotoxic activity of bokkosin, a novel diterpene-substituted chromanyl benzoquinone from *Calliandra portoricensis*. Frontiers in Chemistry. 2020;**8**:574103

[153] Pannequin A, Quetin-Leclercq J, Costa J, Tintaru A, Muselli A. First phytochemical profiling and in-vitro antiprotozoal activity of essential oil and extract of *Plagiochila porelloides*. Molecules. 2023;**28**(2):616

[154] Peniche-Pavía HA, Medrano-Nahuat D, Torres-Tapia LW, Mut-Martín M, García-Miss R, Peraza-Sánchez SR. Metabolites isolated from the rhizomes of *Dorstenia contrajerva* with anti-leishmanial activity. Phytochemistry Letters. 2016;**18**:140-143

[155] Uc-Cachón AH, Chan-Bacab MJ, Lezama-Dávila CM, Isaac-Márquez AP, González-Sánchez AA, Dzul-Beh ÁJ, et al. Anti-Leishmania activity of the Mayan medicinal plant *Thouinia paucidentata* Radlk extracts. Natural Product Research. 2023;**1-5**:3444-3448

[156] Zheoat AM, Alenezi S, Elmahallawy EK, Ungogo MA, Alghamdi AH, Watson DG, et al. Antitrypanosomal and antileishmanial activity of chalcones and flavanones from *Polygonum salicifolium*. Pathogens. 2021;**10**(2):175

[157] Accioly MP, Bevilaqua CM, Rondon FC, de Morais SM, Machado LK, Almeida CA, et al. Leishmanicidal activity in vitro of *Musa paradisiaca* L. and *Spondias mombin* L. fractions. Veterinary Parasitology. 2012;**187**(1-2):79-84